T0092260

Edited by Indrajit Pan, Mohamed Abd Elaziz, and Siddhartha Bhattacharyya

Swarm Intelligence for Cloud Computing

Swarm Intelligence for Cloud Computing

Edited by
Indrajit Pan
Department of Information Technology, RCC Institute of
Information Technology, India

Mohamed Abd Elaziz
Department of Mathematics, Zagazig University, Egypt

Siddhartha Bhattacharyya
Department of Computer Science and Engineering,
CHRIST (Deemed to be University), India

CRC Press
Taylor & Francis Group
Boca Raton London New York

CRC Press is an imprint of the
Taylor & Francis Group, an **Informa** business

A CHAPMAN & HALL BOOK

First edition published 2021
by CRC Press
6000 Broken Sound Parkway NW, Suite 300, Boca Raton, FL 33487-2742

and by CRC Press
2 Park Square, Milton Park, Abingdon, Oxon, OX14 4RN

© 2021 Taylor & Francis Group, LLC

CRC Press is an imprint of Taylor & Francis Group, LLC

Reasonable efforts have been made to publish reliable data and information, but the author and publisher cannot assume responsibility for the validity of all materials or the consequences of their use. The authors and publishers have attempted to trace the copyright holders of all material reproduced in this publication and apologize to copyright holders if permission to publish in this form has not been obtained. If any copyright material has not been acknowledged please write and let us know so we may rectify in any future reprint.

Except as permitted under U.S. Copyright Law, no part of this book may be reprinted, reproduced, transmitted, or utilized in any form by any electronic, mechanical, or other means, now known or hereafter invented, including photocopying, microfilming, and recording, or in any information storage or retrieval system, without written permission from the publishers.

For permission to photocopy or use material electronically from this work, access www.copyright.com or contact the Copyright Clearance Center, Inc. (CCC), 222 Rosewood Drive, Danvers, MA 01923, 978-750-8400. For works that are not available on CCC please contact mpkbookspermissions@tandf.co.uk

Trademark notice: Product or corporate names may be trademarks or registered trademarks, and are used only for identification and explanation without intent to infringe.

ISBN: 978-0-367-03055-1 (hbk)
ISBN: 978-0-429-02058-2 (ebk)

Typeset in Palatino
by codeMantra

Indrajit Pan would like to dedicate this book to his friends of Friend-o-Mania group, Manash Sinharoy, Avijit Das, Palash Kumar Ghosh, David Thomas and Ratan Biswas, for their unconditional support as and when needed as an extended family members and part of an emphatic support system.

Mohamed Abd Elaziz would like to dedicate this book to his family and Rehab Ali for supporting him and giving more motivation in his life.

Siddhartha Bhattacharyya would like to dedicate this book to his father Late Ajit Kumar Bhattacharyya, his mother Late Hashi Bhattacharyya, his beloved wife Rashni and his research scholars Sourav, Sandip, Pankaj, Hrishikesh, Debanjan, Koyel, Alokananda, and Tulika.

Contents

Preface

Cloud computing is a newly developed platform for providing IT services. Like the subscribers of traditional utility services, subscribers of cloud computing pay only for the services and resources that they use. Major IT companies, such as Amazon, Google, and Microsoft, are currently providing cloud services and resources. These companies consider cloud computing as the future of IT operations. Subscribers do not have to build or maintain their own IT infrastructures. Cloud computing allows subscribers to rapidly and inexpensively re-provision IT resources for various needs.

Due to flexibility, security, availability, scalability and affordability, cloud computing has begun to attract the attention of IT stakeholders. The features provided by cloud computing are extremely beneficial for small organizations that lack funding to maintain their own infrastructure. Nowadays, many organizations are leaning toward this new technology for various IT-related infrastructural and application needs. So far, cloud computing has been explored up to an extent of its infrastructural and security-related know how; However, a wide scope remains for the research.

Primitive technological development needs to be augmented with high-end intelligent approaches so that the cloud setup becomes more robust and efficient. People are still concerned about its sustainability and stability of infrastructure and security of services and data over the cloud. Since it is run over a distributed platform, many complexities and constraints have to be considered at practical level implementation.

This book will present some state-of-the-art research involving swarm and hybrid swarm intelligence for cloud security, reliability, and infrastructural stability. Security-related task will focus on public auditing of data and services. Infrastructural development will be focused on various aspects like task scheduling, virtual machine allocation, load balancing and optimization, deadline handling, power-aware profiling, fault resilience, cost effective design, and energy efficiency.

In the last few decades, a novel branch of intelligent computation algorithm has been inspired by a swarm intelligence theory that imitates the behavior of animals. These algorithms are successfully applied to solve many kinds of simple and complex problems in several fields such as optimization problems, pattern recognition, image processing, features section, and task scheduling in cloud and parallel computing. Cloud computing has become the preferred environment for several companies. Moreover, it has helped several companies to overcome many server issues by utilizing their characteristics such as reliability, flexibility, high scalability, and security. Therefore, many intelligent computation algorithms are employed to improve this environment. In Chapter 1, an overview of swarm intelligence for solving the scheduling problems of tasks in cloud computing is presented, including particle swarm optimization, cat optimization algorithm, artificial bee colony, lion optimization algorithm, whale optimization algorithm, Bat algorithm, gray wolf optimizer, cuckoo search algorithm, hybrid swarm algorithms, and multi-objective swarm optimization. All these algorithms are described and presented with their achievements in solving task scheduling issues in cloud computing.

Cloud Computing has revolutionized the way Information Technology (IT) operates. Resource sharing in the cloud enables multiple users or providers to share resources, which improves the utilization of their resources. In the cloud, aggregated resources create a resource pool. The resource pools handle the dynamic demand of IT. Resource sharing enables cloud providers and customers to reduce their capital expenditure, operational expenditure, and total cost of ownership. Multiple cloud service providers connect to form a federation to share their resources among themselves to meet the dynamic demand of their customers. Chapter 2 discusses resource sharing in intercloud, social cloud, Mobile Cloud, Manufacturing Cloud, Vehicular cloud, and Green cloud. In each cloud, various resource sharing approaches, strength and challenges are discussed.

In the cloud computing environment, comparing providers and selecting the best one among them is one of the most important issue/decisions to customers/clients. The difficulty of provider selection decision increases with increasing number of cloud providers and its offering. However, from the customers' perspective, this enables them to take the advantage of getting its requirement from multiple cloud providers and reducing the dependency on single cloud provider. Multi-cloud has been the focus of recent research—a strategy that assists customers to avoid vendor lock-in problem. Chapter 3 presents a general model for selecting providers in a multi-cloud environment, considering any number of IaaS services based on two evaluation criteria: cost and performance. The problem was formulated as integer programming and it was approved to be a NP hard problem. Consequently, to solve it, three metaheuristic algorithms have been used: Genetic Algorithm (GA), Harmony Search (HS) and Particle Swarm Optimization (PSO). In order to test and compare the performance of the proposed algorithms, a case study was generated.

Twenty-first century has observed prolific progress of cloud computing paradigm. Many individuals and organizations are now heavily dependent on cloud services to run their operations and businesses. Many aspects of cloud computing are still in development under intense observation of research community. The underlying requirement of cloud computing is a network that empowers the connection among stake holders and resources. Security of resources, specifically data, is a key concern along with right deployment of resources to deliver optimal performance. Chapter 4 proposes two efficient approaches together toward reliable service execution of cloud. One part provides an ant colony optimization approach for dynamic resource scheduling and the other part ensures secure auditing of data during dynamic resource allocation. Experimental analysis shows performance improvement in terms of request processing time, auditing time, successful resource allocation and average utilization of servers with different volumes of clients among all available servers.

The most critical action in the Internet of Things (IoT), particularly in the cloud computing (CC) ecosystem, is job/task scheduling. The rule of the task scheduling (TS) means that how to arrange/schedule the jobs over the given virtual machines (VMs) by decreasing the Makespan measured (Mm) values and the cost need. Various scholars propose several scheduling algorithms for solving this problem of scheduling the tasks in cloud computing ecosystems. In Chapter 5, a task scheduling method is proposed using a multi-objective design model and the Gray Wolf Optimizer (GWO), called TS-GWO. First, the multi-objective criterion measures the fitness function by determining the cost value of the CPU ratio and memory size (Me). The fitness function is determined by calculating the Makespan value and demand value. The proposed method (TS-GWO) can expertly arrange the given jobs to the offered VMs while keeping the smallest Makespan value and cost value. Conclusively, the performance of the proposed method (TS-GWO) is investigated and matched with the basic GWO for the testing measures: Makespan ration and cost value.

Due to the recent developments witnessed in the world of computing, IT trends are continuously changing. This includes changes in the way information is gathered, managed and processed. In this regard, cloud computing is considered to be the state-of-the-art technology to host and manage stored information on a network of shared resources. With the arrival of SDN technology, the intelligence from hardware devices shifted from vendors to application awareness concepts. This also led to the ease of development of network management functions and features. In Chapter 6, the authors give a quick look at the various aspects of application awareness in data center resource management.

Biologically inspired optimization techniques are applied in various fields for finding a quick optimal or near optimal solution to computationally hard problems. Study on the application of these techniques for task scheduling in cloud computing systems is worth due to the dynamic and unpredictable nature of task arrival. The quality of a task schedule in cloud computing systems depends on the response time that a user of the cloud system experiences and the cost of the service at both cloud service provider's end and at the user side. Optimizing such conflicting factors in a dynamic environment is challenging since it is a multi-objective optimization problem. Many scientific and large-scale applications are characterized by a huge number of homogeneous tasks that can be executed in parallel. Chapter 7 considers the scheduling of such tasks. Three swarm intelligence algorithms namely Particle

Swarm Optimization (PSO) technique, Artificial Bee Colony (ABC) algorithm and Ant Colony Optimization (ACO) algorithm and an evolutionary computation based algorithm namely Genetic Algorithm (GA) are applied to solve the problem. Experiments are done on different types of data sets and a comparative study on the results is performed to determine the suitability of these algorithms in the cloud environment.

This volume explores different research perspectives based on swarm intelligence or hybrid swarm intelligence based techniques to deliver robust and stable mechanism for cloud infrastructural support and cloud security.

India, Egypt
Indrajit Pan
Mohamed Abd Elaziz
Siddhartha Bhattacharyya

Editors

Dr. Indrajit Pan has done his Bachelor of Engineering (BE) in Computer Science and Engineering with Honors from The University of Burdwan in 2005 and completed Master of Technology (M.Tech) in Information Technology from Bengal Engineering and Science University, Shibpur in 2009. He was the recipient of University Medal for his academic performance during Masters. He obtained Ph.D. in Engineering from Department of Information Technology, Indian Institute of Engineering Science and Technology, Shibpur in 2015. His research title was Design and Analysis of Droplet Routing Algorithms in Digital Microfluidic Biochip.

His present research interest includes social network analysis and cloud computing. Indrajit is now serving as Associate Professor of Information Technology Department at RCC Institute of Information Technology, Kolkata with an experience of 14 years in teaching.

Indrajit takes active interest in organizing international conferences, symposiums, seminars and faculty development programs at different capacities. He was the member of different such organizing committees and also served in the capacity of General chair, organizing secretary and program chair at multiple occasions. He has also served as technical program committee members in many international conferences and reviewer in some IEEE transactions and Elsevier journals.

He has around forty research publications in different International journals, edited books and conference proceedings. Three research scholars are enrolled and working under his supervision for their doctoral dissertation.

Indrajit has coauthored five published research volumes with prestigious publishers like John Wiley, CRC Press, De Gruyter and IGI Global. In addition to these he has coauthored four international journal proceedings with IEEE, USA and two International symposium proceedings with Springer Nature. He served as guest editor in *International Journal of Hybrid Intelligence* for special issue on Hybrid computational intelligence in big data analytics and cloud computing.

Indrajit is the Senior member of IEEE, USA and Member of ACM, USA.

Mohamed Abd Elaziz was received the B.S. and M.S. degrees in Computer Science from the Zagazig University, in 2008 and 2011, respectively. He received Ph.D. degree in mathematics and computer science from Zagazig University, Egypt in 2014. From 2008 to 2011, he was Assistant lecturer in Department of Computer Science. He is currently an Associate Professor with the Mathematical Department, Zagazig University. He is the author of more than 100 articles. His research interests include machine learning, signal processing, image processing, cloud computing, Evolutionary algorithms. Mohamed has coauthored two published books.

Siddhartha Bhattacharyya did his Bachelors in Physics, Bachelors in Optics and Optoelectronics and Masters in Optics and Optoelectronics from University of Calcutta, India in 1995, 1998 and 2000 respectively. He completed PhD in Computer Science and Engineering from Jadavpur University, India in 2008. He is the recipient of the University Gold Medal from the University of Calcutta for his Masters. He is the recipient of several coveted awards including the Distinguished HoD Award and Distinguished Professor Award conferred by Computer Society of India, Mumbai Chapter, India in 2017, the Honorary Doctorate Award (D. Litt.) from The University of South America and the South East Asian Regional Computing Confederation (SEARCC) International Digital Award ICT Educator of the Year in 2017. He has been appointed as the ACM Distinguished Speaker for the tenure 2018–2020.

He is currently serving as a Professor in the Department of Computer Science and Engineering of Christ University, Bangalore. He served as the Principal of RCC Institute of Information Technology, Kolkata, India during 2017–2019. He has also served as a Senior Research Scientist in the Faculty of Electrical Engineering and Computer Science of VSB Technical University of Ostrava, Czech Republic (2018–2019). Prior to this, he was the Professor of Information Technology of RCC Institute of Information Technology, Kolkata, India. He served as the Head of the Department from March, 2014 to December, 2016. Prior to this, he was an Associate Professor of Information Technology of RCC Institute of Information Technology, Kolkata, India from 2011 to 2014. Before that, he served as an Assistant Professor in Computer Science and Information Technology of University Institute of Technology,

The University of Burdwan, India from 2005 to 2011. He was a Lecturer in Information Technology of Kalyani Government Engineering College, India during 2001–2005. He is a co-author of 5 books and the co-editor of 54 books and has more than 280 research publications in *International Journals and Conference Proceedings* to his credit. He has got two PCTs to his credit. He has been the member of the organizing and technical program committees of several National and International Conferences. He is the founding Chair of ICCICN 2014, ICRCICN (2015–2018; 2020), ISSIP (2017, 2018; 2020) (Kolkata; Bangalore, India; Zagreb, Croatia), DoSIER (2019; 2020) (Kolkata; Santiniketan, India). He was the General Chair of several international conferences like WCNSSP 2016 (Chiang Mai, Thailand), ICACCP (2017, 2019) (Sikkim, India) and ICICC (2018; 2020) (New Delhi, India) and ICICC 2019 (Ostrava, Czech Republic).

He is the Associate Editor of several reputed journals including Applied Soft Computing, IEEE Access, Evolutionary Intelligence and IET Quantum Communications. He is the editor of *International Journal of Pattern Recognition Research* and the founding Editor in Chief of *International Journal of Hybrid Intelligence, Inderscience.* He has guest edited several issues with several International Journals. He is serving as the Series Editor of IGI Global Book Series Advances in Information Quality and Management (AIQM), De Gruyter Book Series Frontiers in Computational Intelligence (FCI), CRC Press Book Series on Computational Intelligence and Applications, CRC Press Book Series on Quantum Machine Intelligence, Wiley Book Series Intelligent Signal and Data Processing, Elsevier Book Series Hybrid Computational Intelligence for Pattern Analysis and Understanding and Springer Tracts on Human Centered Computing.

His research interests include hybrid intelligence, pattern recognition, multimedia data processing, social networks and quantum computing.

Dr. Bhattacharyya is a life fellow of Optical Society of India (OSI), India, life fellow of International Society of Research and Development (ISRD), UK, a fellow of Institute of Electronics and Telecommunication Engineers (IETE), India and a fellow of Institution of Engineers (IEI), India. He is also a senior member of Institute of Electrical and Electronics Engineers (IEEE), USA, International Institute of Engineering and Technology (IETI), Hong Kong and Association for Computing Machinery (ACM), USA. He is a life member of Cryptology Research Society of India (CRSI), Computer Society of India (CSI), Indian Society for Technical Education (ISTE), Indian Unit for Pattern Recognition and Artificial Intelligence (IUPRAI), Center for Education Growth and Research (CEGR), Integrated Chambers of Commerce and Industry (ICCI), and Association of Leaders and Industries (ALI). He is a member of Institution of Engineering and Technology (IET), UK, International Rough Set Society, International Association for Engineers (IAENG), Hong Kong, Computer Science Teachers Association (CSTA), USA, International Association of Academicians, Scholars, Scientists and Engineers (IAASSE), USA, Institute of Doctors

Engineers and Scientists (IDES), India, The International Society of Service Innovation Professionals (ISSIP) and The Society of Digital Information and Wireless Communications (SDIWC). He is also a certified Chartered Engineer of Institution of Engineers (IEI), India. He is on the Board of Directors of International Institute of Engineering and Technology (IETI), Hong Kong.

Contributors

Aaqif Afzaal Abbasi
Department of Software Engineering
Foundation University
Islamabad, Pakistan

Almas Abbasi
Department of Computer Science
International Islamic University
Islamabad, Pakistan

Hisham M. Abdelsalam
Faculty of Computers and
    Information
Cairo University
Cairo, Egypt

Hayfa Y. Abuaddous
Faculty of Computer Sciences and
    Informatics
Amman Arab University
Amman, Jordan

Laith Abualigah
Faculty of Computer Sciences and
    Informatics
Amman Arab University
Amman, Jordan

Hamzeh Alabool
Department of Information
    Technology
Saudi Electronic University
Riyadh, Saudi Arabia

Yousif A. Alhaj
School of Computer Science and
    Technology
Wuhan University of Technology
Wuhan, China

Mohammed A. A. Al-qaness
School of Computer Science
Wuhan University
Wuhan, China

Mohammad Alshinwan
Faculty of Computer Sciences and
    Informatics
Amman Arab University
Amman, Jordan

A. S. Ajeena Beegom
Department of Computer Science
    and Engineering
College of Engineering Trivandrum
Trivandrum, India

R. V. Belfin
Department of Computer Science
    and Engineering
Karunya Institute of Technology and
    Sciences
Coimbatore, India

Mofleh Al Diabat
Department of Computer Science
Al Albayt University
Mafraq, Jordan

Rania M. ELEraky
Computer Curriculum and Methods
  Department
University of Bisha
Bisha, Saudi Arabia
and
Faculty of Specific Education
Damietta University
Damietta, Egypt

Ahmed A. Ewees
Computer Department
Damietta University
Damietta, Egypt

Ammar Hawbani
School of Computer Science and
  Technology
University of Science and Technology
  of China
Hefei, China

Rehab A. Ibrahim
School of Computer Science and
  Technology
Huazhong University of Science and
  Technology
Wuhan, China

P. Karthikeyan
Department of Computer Science
  and Engineering
Karunya Institute of Technology and
  Sciences
Coimbatore, India

Ahmad M. Khasawneh
Faculty of Computer Sciences and
  Informatics
Amman Arab University
Amman, Jordan

E. Kirubakaran
Department of Computer Science
  and Engineering
Karunya Institute of Technology and
  Sciences
Coimbatore, India

K. Martin Sagayam
Department of Electronics and
  Communication Engineering
Karunya Institute of Technology and
  Sciences
Coimbatore, India

Amany M. Mohamed
Faculty of Computers and
  Information
Cairo University
Cairo, Egypt

Neggaz Nabil
Faculte' des Mathe'matiques et
  Informatique, De'partement
  d'informatique
Universite' des Sciences et de la
  Technologie d'Oran Mohamed
  Boudiaf, Laboratoire Signal Image
  PArole (SIMPA)
Oran, Algrie

Indrajit Pan
Department of Information
  Technology
RCC Institute of Information
  Technology
Kolkata, India

B. L. Radhakrishnan
Department of Computer Science
  and Engineering
Karunya Institute of Technology and
  Sciences
Coimbatore, India

M. S. Rajasree
Department of Computer Science
Government Engineering College
Trivandrum, India

Soumitra Sasmal
Department of Information
   Technology
Techno Main, Salt Lake
Kolkata, India

Mohammad Shehab
Computer Science Department
Aqaba University of Technology
Aqaba, Jordan

S. Sudhakar
Department of Computer Science
   and Engineering
Karunya Institute of Technology and
   Sciences
Coimbatore, India

1

A Survey of Swarm Intelligence for Task Scheduling in Cloud Computing

Ahmed A. Ewees

Damietta University

Neggaz Nabil

Université des Sciences et de la Technologie d'Oran Mohamed Boudiaf, Laboratoire Signal Image PArole (SIMPA)

Rehab A. Ibrahim

Huazhong University of Science and Technology

Rania M. ELEraky

University of Bisha.
Damietta University

CONTENTS

1.1   Introduction

Cloud computing (CC) has become the preferred environment for large-scale applications which need diverse platforms. It is an evolution of grid, cluster, and distributed systems [12]. It has many benefits, such as access to resources on demand; low cost due to shared resources; large storage; application of service level agreement, and use of virtual machines (VMs) to increase the resources' utilization where applications, systems, and infrastructure are completely provided as a service; and customers can access these systems anytime from anywhere [12]. Figure 1.1 shows the main idea of CC [71].

The benefits of CC, in the recent years, have helped several companies to overcome many server-related issues by utilizing their characteristics, such as reliability, flexibility, high scalability, multi-tenancy, security, data flow, quality of service (QoS), and providing better maintenance and support [12,59]. The main goal of CC is to decrease the computation cost for consumers who are using these services [12]. QoS requires companies to provide the desired performance to their customers such as sending jobs to appropriate resources, scheduling, and producing the outputs. All these points should be efficiently performed within a short time with satisfying results [46].

In addition, there are several challenges and issues that should be considered in providing the services to the customers in CC systems, for example, task scheduling, load balancing, security issues, VMs placement, energy consumption, and storage space. However, job or task scheduling is still considered as an essential phase in CC since it works to minimize the cost toward resource utilization.

Job scheduling in CC is one of the well-known optimization problems and plays an essential role in improving reliable and flexible platforms [75]. Job

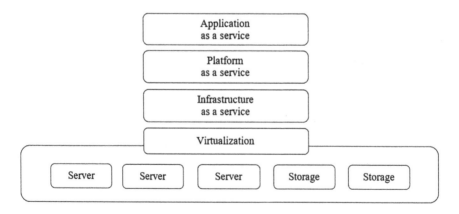

FIGURE 1.1
The main idea of CC [71].

scheduling works to map the resources to jobs. The optimal scheduling method should guarantee a QoS and good performance of the CC by taking into account some criteria, such as computation time and cost [86]. It should maximize the utilization of VMs while minimizing their computation time [75]. Figure 1.2 illustrates a sample scenario of task scheduling problem in CC [59].

Jobs can be dependent or independent. If they have precedence orders, their scheduling process is called dependent or workflow scheduling, otherwise it is called independent scheduling [46]. The diversity of CC's resources and the varying types of jobs lead to make the scheduling method an NP-hard problem [86]. VM should process the jobs as early as possible, but because of the huge numbers of jobs and VMs in CC, it may not schedule the jobs to resources optimally. Therefore, efficient job-scheduling methods are required to handle this issue [29]. There is no optimal method to solve all kinds of CC's issues. Exhaustive search methods are not efficient as they consume high costs in generating schedules. In contrast metaheuristics methods handle these issues by producing near-optimal solutions in a sensible time. However, there are many techniques used to handle these problems, but the metaheuristics optimization techniques are still an effective method to deal with this challenge [46]. Swarm intelligence is a type of metaheuristics optimization method that simulates the natural behavior of insects, animals, or birds in swarming, preying, or flocking such as Bat algorithm [98], particle swarm optimization (PSO) [25], whale optimization algorithm (WOA) [65], gray wolf optimization (GWO) [66], and artificial bee colony (ABC) [48].

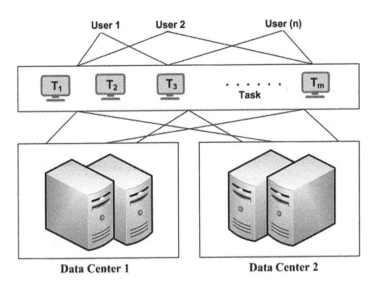

FIGURE 1.2
A sample scenario of a task scheduling problem in CC [59].

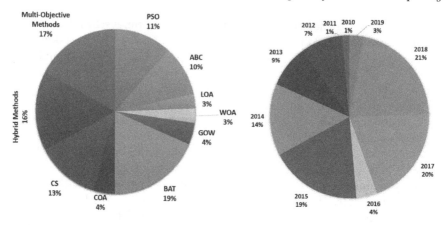

FIGURE 1.3
The distribution of the algorithms in this paper.

These methods are applied successfully in several fields because of their effectiveness and efficiency in dealing with complex and large problems such as image processing [26,33,34,40], features selection [28,31], and prediction problems [4–6,30,81].

In this chapter, we introduce a review of swarm intelligence applied for CC specially in tasks/jobs scheduling based on metaheuristic methods, namely PSO, lion optimization algorithm (LOA), ABC, GWO, Bat algorithm (BAT), cat swarm optimization (CSO), WOA, cuckoo search algorithm (CSA), hybrid algorithms, and multi-objective algorithms based on these methods. We reviewed more than 70 studies published from 2010 to 2019 in this field. Figure 1.3 summarizes these methods and their distribution over the years. From this figure, we can notice that the hybrid and multi-objective algorithms are the most popular methods in solving CC issues and the efforts to use these methods were increased in 2017 and 2018. Therefore, we predict that the use of these methods will also increase in 2019.

The rest of the sections of this chapter are arranged as follows. Section 1.2 shows the problem definition of task/job scheduling in CC. Section 1.3 presents a review of swarm intelligence methods as well as their hybridization and multi-objective versions to solve task scheduling issues in CC. Section 1.4 concludes the paper.

1.2   Task Scheduling Problem Formulation

In this section, the overview of a task scheduling problem in CC is presented. A task scheduling issue can occur if a program has N tasks, $T = t_1, t_2, \ldots, t_N$, each of which needs to be assigned to a VM to be processed; and there are VMs in the environment (VMs = VM_1, VM_2, ..., VM_M; M is the total number of VMs); whereas, this assignment requires times, called computation time, to be completed. Therefore, each task has an arrival time and firm end time. Due to heterogeneity of VMs, each VM has different cost of usage, different computation speed, and limited capacity; therefore, the cost and computation time of tasks are different. Equations 1.1 and 1.2 represent the matrix of the computation time (ct) of task (t_n) and its cost (cost), respectively, as follows [21]:

$$CT = (ct_{nm})_{(N \times M)} \tag{1.1}$$

where, ct_{nm} denotes the computation time of task (t_n) on VM_m machine.

$$\text{Cost} = (\text{cost}_{nm})_{(N \times M)} \tag{1.2}$$

where, cost_{nm} is the cost usage of task (t_n) on VM_m machine.

The aim of task scheduling is to allocate the n tasks to the appropriate m VMs to minimize the makespan of the task as well as balance the loads on CC machines. Thus, Equations 1.3 and 1.4 are used to calculate the total computation time and costs for n tasks, respectively [21]:

$$CT_{\text{total}} = \Sigma_N n = 1 CT_{nm} \tag{1.3}$$

$$\text{Cost}_{\text{total}} = \Sigma_{n=1}^{N} \text{Cost}_{nm} \tag{1.4}$$

In addition, the task scheduling method should consider the load balance to effectively allocate tasks. Equation 1.4 calculates the load balance Load_{bal} of a VM.

$$\text{Load}_{\text{bal}} = \Sigma_{n=1}^{M} |b_m - \bar{b}| \tag{1.5}$$

where b_m indicates the load of VM_m and $\bar{b}$ represents the mean load of all VMs. Consequently, the task scheduling problem can be described as follows in Equation 1.61.6:

$$\text{Minimize}(CT_{\text{total}}, \text{Cost}_{\text{total}}, \text{Load}_{\text{bal}}) \tag{1.6}$$

1.3 The Swarm Intelligence Application for Cloud Computing

Swarm intelligence methods vary from simple to complex. Such methods are applied to obtain a best solution in a small computation time. Therefore, various swarm intelligence methods for task scheduling in CC can be used. This section illustrates an overview of the most applied swarm intelligence methods for scheduling tasks in the cloud, such as PSO, ABC algorithm, LOA, WOA, CSA, GWO, cat optimization algorithm, Bat algorithm, hybrid swarm algorithms, and multi-objective swarm optimization.

1.3.1   Particle Swarm Optimization

The PSO was proposed by the authors of [25]. It was designed to simulate the communication behavior of the bird's group in nature to share their knowledge in migrating, flocking, and hunting. The bird group is called swarm and each member is called particle. Therefore, a swarm consists of a set of particles which denote a solution. The particle's position is changed based on its own experience and neighbors. In the mathematical model, a swarm begins by generating random particles, each of which has a position and a velocity in the solutions dimension.

These values are updated in each loop to find the best position by applying a fitness function to maintain the solution and try to move it toward the optimal value of the given problem. This sequence is repeated until meeting the desired solution. In the job scheduling issue, the particles are symbolized as $(x \times y)$, where y indicates the tasks' numbers and x denotes the resources' number to perform the jobs scheduling issue. Figure 1.4 shows the flowchart of PSO [91].

PSO is widely used to solve jobs scheduling problem in CC. In Ref. [43], the authors used a modified PSO for solving the length of average scheduling and successful execution ratio in a cloud. Different numbers of tasks were used, namely 100, 300, 500, 700, and 900. The experiments were performed using CloudSim platform and they showed better results than other methods. PSO was also applied in task scheduling in CC [57]. The PSO was improved by adaptive weights to change the weight with increasing the iterations' number and produce random weights in the next stage to avoid PSO from getting trapped in local optima. The CloudSim platform was used. The authors used different numbers of tasks, namely 100, 200, and 500 as well as 10 VMs.

In the same trend, PSO was improved in [97]; however, only 6 VMs and 15 tasks were tested using CloudSim platform. In addition, the authors of [21] reported that they effectively applied PSO in job scheduling to obtain better load balancing and minimize the ratio of deadline missing. The experiment used 60 VMs and 250, 300, 350, 400, and 450 tasks. In Ref. [13] the PSO was used to reduce the makespan and cost simultaneously. It used 10 VMs and 99 tasks, but the experiment environment was not mentioned. In Ref. [7] PSO improved based on a ranking strategy and multi-objective concept to schedule tasks to VMs in order to maximize the throughput of the system and minimize the waiting time. The experiment used 500–2,000 VMs and 500–4,000 tasks. The authors reported that the throughput was improved up to 40% and the waiting time was reduced to 30%.

Another effort in solving job scheduling issue by PSO is presented in [76]. This study suggested transferring only overloaded tasks from VM to decrease the downtime in the migration process. The experiment was performed on the CloudSim platform and 5 VMs and 10 cloudlets (as overloaded VM) were applied. The results showed that the study method could reduce the task scheduling time than traditional methods and increased the QoS by

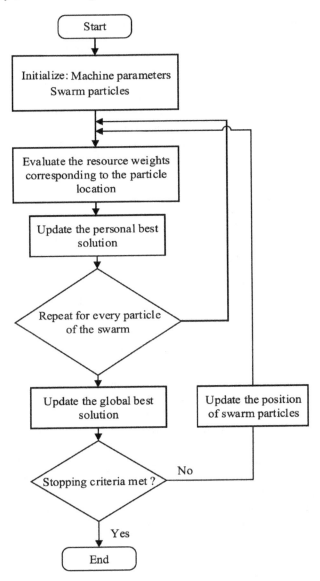

FIGURE 1.4
The flowchart of the PSO [91].

cloud clients. In [19], the PSO was compared with Naïve algorithm to schedule the tasks in CC in smallest computation. The experiment was performed on the CloudSim platform and 5 VMs and 100 cloudlets were applied. The results showed that the PSO had the smallest computation time, whereas the naïve algorithm was computationally expensive when applying on large datasets.

1.3.2   Artificial Bee Colony Algorithm

The ABC algorithm is a type of swarm intelligence method; it mimics the foraging behavior of the colony of real honey bees. ABC algorithm was proposed in 2005 by Karaboga [48]. ABC algorithm consists of three types of bees' groups:(1) the worker bees group which search for food's sources near the recently visited food sources; (2) the onlooker bees group which work to find a food source using the information provided by the first group; and (3) the scout bees group search randomly to discover more food locations with a high amount of nectar. In an optimization method, the solution is represented as a food source and the solution's quality is represented by the amount of nectar [48]. Figure 1.5 illustrates the flowchart of the ABC algorithm [71].

The ABC algorithm was applied to deal with several CC optimization problems; some of these works are presented in the following. The ABC algorithm in [83] was improved and utilized to solve task scheduling problem in CC to map tasks to available resources. It worked to improve the computation performance and the resource cost by minimizing execution time and costs. In addition, it grouped the tasks and then sent them to the cloud resources. The experiments were performed using the CloudSim platform to simulate cloud's platform. Four cloudlets were examined (i.e., 25, 50, 75, and 100). The results of the ABC algorithm outperformed activity-based costing algorithm. Another effort to solve the problem of task scheduling has been found in [71], in which the system's production was enhanced by scheduling tasks efficiently to VMs using the ABC algorithm. The experiments were performed using the CloudSim platform. The experiments tested a different number of VMs (from 20 to 100) and a different number of cloudlets (from 100 to 900); however, its results were not compared with the state-of-the-art algorithms.

The authors of [10] presented a version of the ABC method in order to maintain load balancing and migration of tasks across VMs as well as minimize the makespan. The food sources in this algorithm were represented as underloaded machines. The experiments showed the superiority of the proposed method and were performed using different numbers of cloudlets (i.e., 10, 15, 20, 25, and 30); then the output makespan and the number of jobs migration between the ABC and the modified ABC were compared. Similarly, the ABC algorithm was applied in [52] to improve load balancing and tasks migration through VMs. The overload tasks that were deleted from the VMs were represented as honey bees and the priority of these tasks was taken into account. The experiments tested 10 to 40 tasks and calculated using the CloudSim platform. The results showed an improvement in computation time and minimizing the response time of VMs. In the same trend, the authors of [85] maintain load balancing and migration of tasks using different numbers of tasks, namely 20, 30, 40, 50, and 60. The experiments showed a cost reduction of using VM instances. Another version of the ABC algorithm was proposed by the authors of [58]. It combined the ABC and ant colony optimization (ACO) algorithms to solve task scheduling in CC. The experiments were applied using

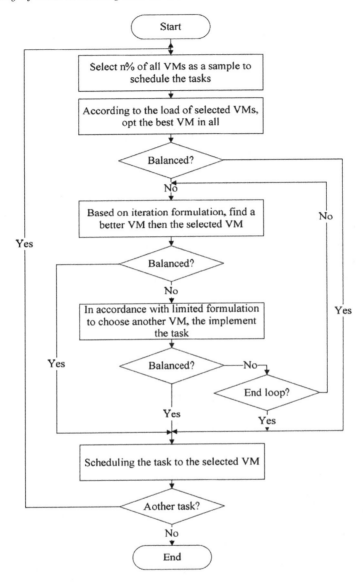

FIGURE 1.5
The ABC flowchart [71].

the CloudSim platform and six cloudlets were used (i.e., 10, 100, 200, 300, 500, and 1,000). The proposed method showed improvement equals 11% than the classical version of the ABC algorithm.

The ABC algorithm was also combined with the GA algorithm to perform energy-efficient task scheduling [53]. The experiments were applied over the CloudSim platform and the number of tasks and VMs varied between 10 and 40. The results depicted that the proposed method showed good performance in energy consumption and minimum makespan than the GA algorithm.

1.3.3   Lion Optimization Algorithm

The LOA is an optimization method which was developed in 2015 by the authors of [101]. It aims to simulate the behavior and lifestyle of lions, based on hunting, mating, and defense [101]. The lion's Kingdom is composed of two kinds of life organisms known as resident and nomads. The term "Pride" represents a set of residents who live together. We note that the residents can switch their lifestyle (i.e., the resident can become a nomad and vice versa). The hunting principle used for lions is specific because they seek the prey by cooperation. The hunting concept consists to assign three roles for lioness as left, center, and right wing as shown in Figure 1.6. During the chase, each lioness updates its location based on their placement and the locations

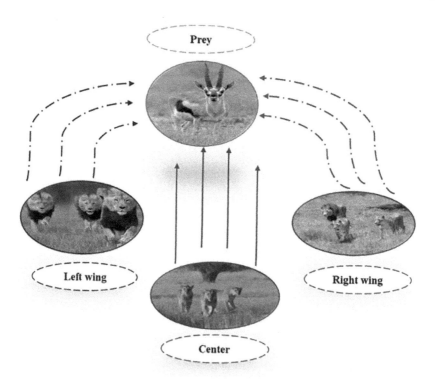

FIGURE 1.6
The natural behavior of lions hunting.

of all elements. The lionesses run very quickly in an exponential way which facilitates the task of hunting prey so that the lionesses work together to catch the victim. For this, they encircle the prey and quickly attack the victim from the opposite position. Due to this reason, the operator of opposition-based learning (OBL) is used in artificial lion algorithm [101].

The LOA is based on the following three important operators [101]:

- Mating: the purpose of this operator consists to breed new descendants using crossover and mutation. Lionesses can proliferate with one or several resident males while the nomad female shares their desire with a single male that are chosen arbitrary. The process of mating is considered as a linear crossover between parents for generating two descendants.

- Territorial defense: this operator consists to estimate the strength of resident lion and nomad lion according to their fitness. If the fitness of the nomad lion is better than the fitness of the resident lion, then we accept this nomad in the pride and the resident must migrate to another area.

- Territorial takeover: in the natural process, the old territorial males and the new territorial males fight among themselves and the strongest remain in the pride. In artificial life of the resident lions (males and cubs), we evaluate all resident lions based on their fitness and the weakest male leave the troop and emerge as nomad while the rest of the males be converted to resident males.

The flowchart of LOA is described in Figure 1.7.

In Ref. [8], the authors aim to order the jobs by LOA in CC. The goal of this study consists to reduce the makespan of the agent that indicates the objective value for each individual. We initialized randomly the population of lions by devoting cloud jobs to VMs where (%N) is the rate of nomads and the rest of the population is residents which is randomly divided through P subgroups appealed Prides. From this subset, we select S portion of females and the rest is considered as males. During the search process, each pride marks its best-visited position well known as a territory. Furthermore, the strategy of hunting consists to divide the lionesses into seven stalking roles, grouping by left wing, center, and right wing position. During the hunting, the OBL is utilized for encircling and attacking the prey. Other operators are used in LOA such as Mating, which produces new offspring, defenses territorial which allow the emergence of new mature males and the weakest males become nomads. Finally, the concept of migration process forces the nomads to leave the kingdom. Devagnanam and Elango [23] proposed an algorithm called E-LOA for allocating optimal resources in the cloud. This paper combines the LOA with an exponential weighted moving average for providing the optimal solution.

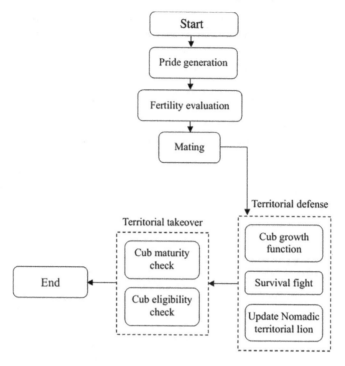

FIGURE 1.7
The flowchart of lion optimization algorithm [73].

1.3.4 Whale Optimization Algorithm

The WOA is an optimization algorithm developed by the authors of [65] in 2016. The WOA simulates the hunting process of the whales that live either in packs or individually. Baleen whales have their own behavior in hunting prey (usually a school of krill or small fish), by generating different bubbles in a circular shape. This method is called bubble-net feeding [96]. The hunting process starts when a whale recognizes the prey's location (the current best solution), then encircles it by creating water balls in a spiral way surrounding the target and swims near to the apparent area. In the WOA, when the near-optimal solution (prey) is reached, other whales try to modify their locations considering the position of this hunt among a shrinking surrounding and across a spiral-form concomitantly, depending on a half percent of probability to select among these two processes. Figure 1.8 illustrates the flowchart of the WOA.

Sreenu and Sreelatha [89] proposed a technique for scheduling tasks in CC known as W-scheduler, based on the WOA and the multi-objective model which is calculated by cost and makespan. Another enhanced version of the scheduling-based WOA are presented in the literature; for example, we found the work of Abdel-Basset et al. who merged the local method with the WOA

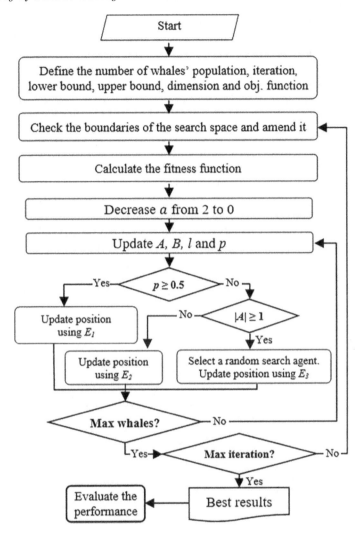

FIGURE 1.8
The flowchart of the WOA [27].

for solving the configuration of schedule flow shop [1]. Recently, the emergence of chaotic theory attracts many researchers in the field of metaheuristics for finding a solution of workflow scheduling in computational cloud.

1.3.5   Gray Wolf Optimization

In 2013, Mirjalili et al., introduced a swarm intelligence technique called GWO [66]. Their idea is based on mimicking the hunting behavior of wolves and the leadership hierarchy. For simulating the leadership hierarchy, we distinguish mainly four categories of gray wolves. Furthermore, the major steps of hunting are focused on tracking, encircling, hunting, and attacking the prey. For applying this algorithm, a random population of wolves is generated wherein each wolf is represented as a vector in N dimension search space. According to their fitness, the four best wolves are assigned to the alpha, beta, and delta groups. For each generation, the location of wolves is changed and the process will be executed for a maximum iterations' number. Figure 1.9 illustrates the flowchart of the GWO.

In task scheduling, the GWO is widely used as an example. Khalili & Belbamir proposed a version named Pareto-GWO for solving dependent tasks of workflow scheduling using multi-objective algorithm [51]. As fitness functions, the authors seek to minimize makespan and cost, and maximize the throughput of a provider's resources. In task scheduling, each solution corresponds to task-resource mapping while GWO is developed for continuous optimization problem whereas the task-resource mapping is a discrete problem. The authors suggest applying the smallest position value propriety for seeking the position of wolves. Natesana and Arun proposed an enhanced version of GWO called mean GWO for obtaining a solution of task scheduling in

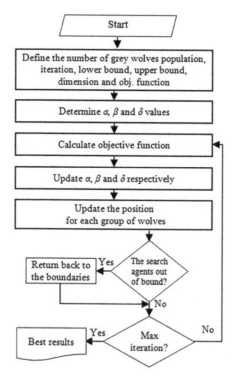

FIGURE 1.9
The flowchart of the GWO [32].

heterogeneous cloud [68]. The authors used energy consumption and makespan as a fitness function. Learning-based opposition is integrated into GWO for parallel scheduling in a cloud platform [67].

1.3.6 Bat Algorithm

Bats are considered mammals, but not much is known about them. However, they have fascinated the greatest researchers and are now of interest to scientists. Their ability to fly, unique in mammals, allow them to devolve at night avoiding obstacles, even at high speed, thanks to their sophisticated echolocation system. They are one of the animals with a very high number of species; there are about 996 species of bats. Their sizes vary between the so-called microchiropter or "microbat" weighing about 1.5–2 g up to the giant bat, called megachiropter or "megabat" which has 2 m of wingspan and weighs over 1 kg. Microchiroptera have a body length ranging from 2 to 11 cm. Most bats use echolocation to some degree among all species. The majority of microchiroptera are insectivores. They emit very loud pulsations through the mouth or the nose (ultrasound). As soon as this ultrasound observes a hurdle (plant, prey), it bounces toward the bat. Their brain can calculate the location, velocity, shape, and the distance of the target to be detected [38]. This principle of bats has helped many scientists to create an algorithm called the Bat algorithm which could be noted as BA. It is considered as a type of the population-based metaheuristic algorithm. Xin-She in 2010 proposed for some optimization problems processing [98]. This is performed through the simulation of the echolocation attitude of bats. Echolocation is a biological sonar (because of lack of vision) that can detect distance, and they also have the ability to distinguish the difference between food/prey and obstacles.

Bats whiffle with a random way of possessing a location, certain frequency, certain speed, nuisance, and a rate for emitting the pulse to search victim. Updating bat velocities and positions are similar to the normal procedure for PSO algorithm. BA can be recognized as a hybrid of the PSO and the exhaustive algorithm of the local search depended on the rate of emitting the pulse and nuisance [41]. The BA is applied for resource scheduling in CC by Jacob for minimizing the makespan, and the author has found that BA showed its efficiency in terms of accuracy than the genetic algorithm (GA) [42]. A combination between the BA and the algorithm of the gravitational scheduling was introduced for task scheduling in CC with taking into account some constraints called deadline ones, in addition to a model called trust model [55].

In Ref. [54], the authors developed a hybrid algorithm between the BA and the harmony search called BATS-HS for processing some problems of the scheduling through the CC. Where selection of a resource depended on the trust; by using the proposed hybrid algorithm the time of computation for the functions or tasks was reduced by about 10.4% versus the selection of a resource in a random way by using the BA. Bat algorithm is further enhanced by chaos theory for the process of scheduling for some functions of tasks through the CC [35]. Bi-objective measures are used for evaluating the performance of the Bat algorithm such as energy consumption and makespan. A flowchart for the BA algorithm is displayed in Figure 1.10 [102].

Raghavan et al. [72] suggested a method that was used as a binary version of the Bat algorithm for scheduling work process in computing the clouds. The mapping of functions or tasks and resources was applied using this method. The proposed method helped in selecting the optimal resources, where the

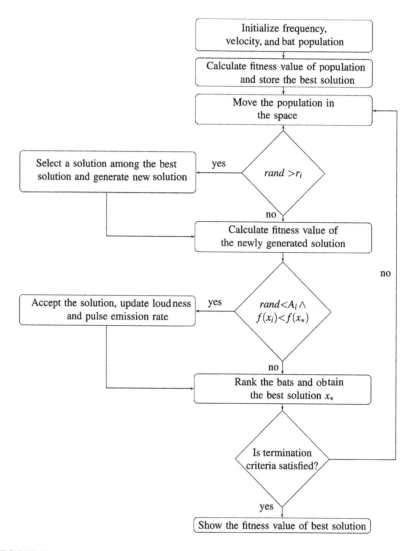

FIGURE 1.10
The flowchart of the BA algorithm [102].

cost of the workflow of this method was the minimum against the compared methods. Marichelvam et al. [62] explored the scheduling problems of the hybrid flow shop (HFS) for multistage using the suggested technique. The problem of the HFS can be considered as the general case of flow shop where there are several machines.

HFS can enter in many industries, such as chemical; textile; and ceramic, iron, and steel, and it can be recognized as an NP-hard problem. The authors improved the BA to be used for solving the HFS problems. They tested their proposed technique on some benchmark functions. The authors compared the proposed Bat algorithm with some popular algorithms such as the GA in addition to the PSO, and the proposed method was proved to have the superiority. Sathya et al. [82] dealt with the process of controlling the load frequency through utilizing the BA depended on the dual mode through scheduling of the controllers of type PI for the power systems that are interconnected. Wang et al. [95] proposed using the Bat algorithm for developing the revenue and energy-aware, in addition to developing combination scheduling method through data center of cloud. The author applied the BA depended on the controller of type dual mode PI for the power system which can be classified as interconnected thermal and multi-area for the parameter PI controllers tuning. The authors tried to compare their controller with some others from the traditional controllers of type PI and their scheduling with the fuzzy gain. The experimental results declared that the proposed BA, which used the scheduling through the dual mode gain of controllers of type PI (BIDPI), provided a suitable transient. The results showed that the new controller has a lower sensitivity in the system parameters changes.

Sagnika et al. [80] presented a solution for the workflow or the data-intensive problem using the BA algorithm through scheduling through the system of the computation of clouds. The proposed algorithm showed a superiority compared with the PSO and CSO algorithms. Such technique gave a cost, with the best convergence, to be the optimal and with suitable distribution. Chaturvedi et al. [20] applied a developed version of the BA to deal with scheduling of workflow in case of the multi-objectives through some clouds which leads to reduce the time of the execution and increasing the reliability through maintaining the budget through a limit of user-specified. This technique was compared with the randomized evolutionary algorithm (BREA), where it depends on using the method of greedy for allocating resources to be suitable for tasks of the workflow, high reliability with low time execution. The experiments proved the superiority of the BA algorithm during its comparison with other evolutionary algorithms. Rani and Kannan [78] focused on the optimal utilization of VMs, modeled around directed acyclic graph with topological ordering to schedule task using nature-inspired Bat algorithm. It can minimize the time waiting of a task and the idle time of VMs due to its high convergence. The authors gave results of the comparison of the Bat algorithm through MATLAB® with First-In-First-out (FIFO), PSO, and Harmonic Search (HS), and it proved the superiority.

Aguirre et al. [3] used the BA to be utilized in the scheduling problems and allocation of work through a multiprocessor, where the makespan and tardiness were the need for the minimization. There is a scheduler for dividing the processors' time among the processes needed to be executed. One task can be executed by a processor, and the scheduler can determine the task that will be in the next process, and the task without restriction of dependence. Malakooti et al. [60] used the Bat algorithm in solving the scheduling problems of the multiprocessor that is energy-aware. The authors performed a comparison between the BA and the GA which proved an efficiency in solving the scheduling issue for multiprocessors and with the objective which is single and its fitness function may be the consumption of energy, or tardiness, or makespan alone or as combination of them for solving the bi-objective multiprocessor or tri-objective multiprocessor through the combination of three of them, energy with makespan, and energy with tardiness. The Normalized Weighted Additive Utility Function can be an effective alternative. The experimental results clarified that the proposed algorithm can find solutions that correspond to the assigned weights in an efficient way.

1.3.7   Cat Swarm Optimization

The cats' behavior has interested the engineers and scientist, hence the birth of an algorithm called "cats optimization algorithm (COA)" introduced by Chu and Tsai [22]. This algorithm is inspired by swarm theory that was applied in the continuous domain of optimization. Two processes are used to understand the behavior of natural cats known as global and local search. In a global search, the seeking mode is used, whereas the local search employed tracing mode. During resting, the cats are highly alert and they react slowly toward their environment; this search is known as seeking mode. As soon as the cats hear the presence of a prey, they react quickly and attack to reach the target; this research is known as tracing mode. In the CSO algorithm, each artificial cat has a position vector and their velocities are updated during the course of the iteration. An active bit is employed for determining the life mode of cat based on their fitness. A simple flowchart for the COA is shown in Figure 1.11 [22].

The authors of [22] proved that the CSO provided better convergence than PSO. Several contribution-based CSOs are developed for a jobs and workflow scheduling. For example, the authors of [15] presented a CSO-based optimization scheduling in order to devote the jobs of an application onto unoccupied devices. The proposed algorithm takes into consideration the cost of the data transfer among the two dependent devices and the cost of jobs on various machines. They tested the proposed CSO using a hypothetical workflow also and made a comparison between the results of the workflow scheduling with the PSO. Their evaluation proved that, firstly, the CSO reached an optimal task-to-resource (TOR) which leads to reduce the total cost. Secondly, the

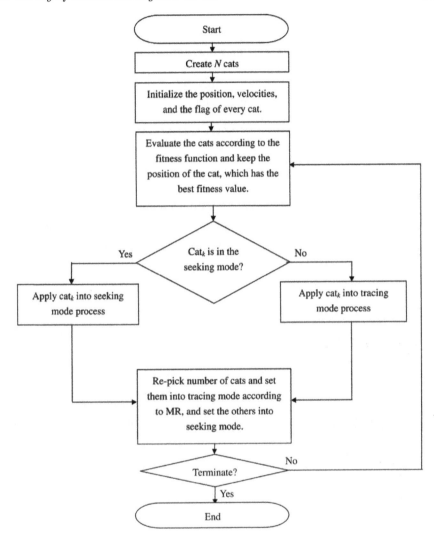

FIGURE 1.11
The flowchart of the cat optimization algorithm [22].

CSO had an advantage over the existing PSO through iterations, finally, the CSO assumed a fair load distribution on the available resources.

Gabi et al. [36] presented an optimization technique known as OTB-CSO, in order to schedule jobs and to reduce the complete time processing using the orthogonal Taguchi (OT) with CSO. The aim of that study was optimizing the task scheduling based on the proposed approach for reducing the total task execution delay. They used the Taguchi orthogonal approach for the resting mode of CSO in order to devote jobs on VMs which have a reduced time of execution. Also, they used the tool of the CloudSim platform for implementing the proposed algorithm whose impact was tested with 5, 10, and 20 VMs with input tasks and performed by makespan and the measures of the degree of imbalance. For 20 VMs, the proposed approach, OTB-CSO, reduced the makespan of the overall scheduled jobs with 2.6% enhancement closed minimum and maximum jobs first. In addition, the proposed method was compared with PSO-LDIW and HPSO-SA based on the degree of imbalance and the obtained result showed an enhancement over existing techniques. Also, it was concluded that the OTB-CSO has the superiority for optimizing task scheduling and enhancing the performance of CC by reducing the task delayed of execution, and a better utilization of the system.

In Ref. [84], they proposed CSO for load distribution of workloads in computational cloud. The principal goal of this study was to achieve higher customer satisfaction and resource utilization ratio. The obtained results were evaluated using a load balancing in a large-scale cloud and showed high performance of computing compared to other algorithms.

1.3.8   Cuckoo Search Algorithm

Yang and Deb [99] suggested the Cuckoo Search (CS) technique, which takes its inspiration from the cuckoos reproduction strategy, who they lay their eggs within a nest, and for increasing the probability of hatching their own eggs, they try to hide the eggs of the others too. The flowchart of the CS algorithm is presented in Figure 1.12 [14].

Many researchers used this technique for scheduling of tasks in the CC. Agarwal and Srivastava [2] suggested using the CS for dealing with the problems of scheduling of tasks, where this can lead to dividing the tasks among the VMs in an efficient manner, in addition to maintaining the minimum of the response time totally. This technique determines the tasks for the VMs based on the processing power, that means that instructions which are almost million per seconds and the tasks' length. They compared the proposed CS approach with the FIFO and greedy algorithms using the simulator of the CloudSim platform. The experiments of this paper clarified that the CS proved to be more efficient than the other algorithms.

Amtade and Miyamoto [9] used the CC to evaluate the CS algorithm against GA. The experimental results showed that the CS proved to be more efficient than the GA according to the objective value.

Mandal and Acharyya [61] proposed swarm intelligence and metaheuristic methods like firefly algorithm (FA), simulated annealing (SA), and CS; such techniques were used for reaching an optimal solution for the scheduling of tasks, and reaching the best resources to try to reduce the total time of processing the VMs. They focused on making the shared resources to be more efficient. The objective of such techniques is reducing the time of processing

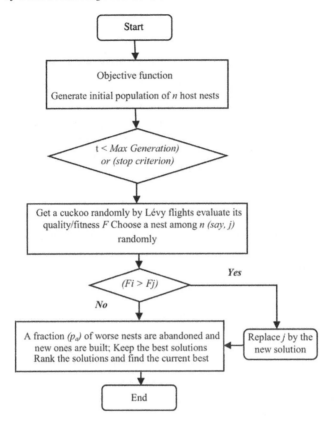

FIGURE 1.12
The flowchart of CS [14].

the VMs which can perform a set of tasks. The results showed that the FA outperforms better than the SA and CS Algorithm.

Marichelvam et al. [63] supposed a version of the CS algorithm for real flow shop scheduling problems as an industrial hybrid by reducing the HFS scheduling problems makespan. They considered the hybrid flow shop in case of the multistage (HFS) scheduling problems. Such problems are recorded as NP-hard. In addition CS algorithm was used for reducing the makespan. Another heuristic algorithm, is the Nawaz-Enscore-Ham (NEH) algorithm that can be associated through the initial solutions for trying to reach the optimal solutions more quickly in the enhanced algorithm. Where the enhanced algorithm was built based on the data from a company of leading furniture manufacturing. The experiments proved that the improved algorithm outperforms the others.

Wang et al. [94] presented an algorithm called NCS, for the scheduling problems of flow shop (FSSP). The smallest position value rule was applied for dealing with the discrete variables and convert continuous solutions into ones. Also, a method based on the NEH algorithm was applied for initializing populations to generate better initial solutions. The experimental study was implemented on a group of Taillard's benchmarks. These experiments proved that the NCS outperformed the standard CS. Burnwal and Deb [11] proposed using the CS for scheduling jobs of a system through reducing the cost of the penalty resulted from the manufacturing delay and the maximization of the time for using the machine. This technique was enhanced using the Lévy flight operator due to which the solution had a discrete nature on a scheduling FMS problem which had 16 machines with 43 jobs. MATLAB was used to perform all experiments. The results were compared to the PSO and GA. The CS showed good performance than the other algorithms.

Ouaarab et al. [70] presented the Discrete Cuckoo Search (DCS) for dealing with the scheduling problems of the Job Shop (JSSP), where they used the Lévy flight operator for enhancing the DCS algorithm through three stages to complete the search space. It applied the local and the global random walk and showed that the DCS can be checked in its performance within benchmarks' set of JSSP from the library named OR. The authors also tested it by comparing it with the PSO algorithm for proving that the DCS had the superiority than the PSO.

Li and Yin [56] used CS based with the memetic algorithm as a hybrid denoted as HCS, for the problem of the permutation flow shop (PFSSP). Also, the authors used a rule of the largest-ranked-value (LRV) which depends on the random key for the CS be suitable for the PFSSP, for converting the continuous location of the CS to a permutation of discrete job. Then hybridized the NEH with the population generated randomly with a specific quality and diversity. Then, the CS was used for including the nest vectors for the process of exploration and improving the capability of the local exploitation. The authors tested the proposed approach performance for solving the PFSSP using the benchmarks, and this proved effectiveness and superiority of the proposed approach.

Hanoun et al. [39] suggested the CS approach for solving the scheduling problem of the job shop in the multi-objective which is single with minimizing the waste of the material and the tardiness time totally. The waste in the material was determined according to some factors to show the material reduction that van be done when producing two jobs through using same materials in a sequence way. There was an estimated factor for computing the cost of the saved processes for each job according to its material type. A formulation of multi-objective optimization problems was adapted to generate the set of schedules that maximize the cost savings and minimize the computation time; all trade-offs were considered for the fitness functions which were conflicting each other. This was performed with developing the CS by using Pareto for a Pareto Archived Multi-Objective Cuckoo Search (PAMOCS) to search for the nondominated optimal solutions of type Pareto. The accuracy of the PAMOCS performance was displayed through the closeness comparison for the resulted solutions of the true Pareto front of the method of enumeration. Experimental results assumed the superiority of the CS for solving such types of problems.

1.3.9 Hybrid Swarm Algorithm

In this context, we distinguish two types of hybridization, the first type consists of combining the global approach with local method known as memetic algorithm (MA), in view to enhance the convergence criteria, whereas the second type combines two or more approaches to global optimization known as cooperative metaheuristics in order to give more flexibility for exploitation and variety of the swarm. The concept of MA consists to apply firstly the global optimization method on the population which are initialized randomly and then applying the local approach on the best individual. For local approach, we found different approaches such as SA, Nelder and Mead's (NM), and tabu search (TS). The efficiency of the MA gave the opportunity to the researchers to test them in real field of application such as the scheduling of jobs and workflow scheduling in cloud.

In Ref. [103], the authors presented a memetic algorithm which combines PSO and SA for reducing the schedule length. The PSO was applied first and then SA was applied for the purpose of improving the convergence. In Ref. [90], the authors described a combination based on PSO and Tabu search for task scheduling. The proposed method consists to split the particles into two groups: the first one updates the position and velocity of particles using global search PSO, whereas the second group updates it using local search (TS) and then we combined both groups in order to find the optimal solution.

Navimipour et al. [69] solved the problem of jobs scheduling using the technique of CS. Their paper aimed to devote the jobs in CC based on the combination between CS and the lévy flight operator. The paper realized three experiments according to the variation of the probability (Pa) from 0.2 to 0.6. They concluded that the lower value of Pa provided high performance in terms of the convergence and the computation time. The authors of [64] tried to deal the problem of energy in jobs scheduling. As solution, Mezmaz et al investigated a parallel method based on GA and DVS (dynamic voltage scaling) for reducing the makespan and the energy. The experimental results showed that the proposed technique accelerated the process of convergence and the rate of energy consumption was enhanced by 47.5% as compared to other algorithms, whereas the makespan was ameliorated by 12% in contrast to other works of the literature. Raju et al. [74] proposed a hybrid algorithm that merged the advantage of ACO and CS, for dealing the problem of task scheduling. In order to allocate the adequate resources, this hybrid algorithm decreased the makespan, since the tasks had been performed within the specified time.

Babukartik and Dhavachelvan [11] developed a method for job scheduling that combined the merits of ACO and CS. Kanagaraj et al. [47] presented an efficient hybrid algorithm based on two global search known as GA and

CS. Their paper aimed to solve the global optimization benchmarks and three design-constrained problems The authors compared the proposed hybrid with other recent ones and used some effective measures for evaluating the performance, from whom, the superiority of the proposed one came. The authors of [93] proposed two functions known as ranging and tuning functions based on PSO for solving the inertia weight. For avoiding the problem of convergence of PSO, the authors aimed to combine BA with PSO for finding the key to the scheduling jobs in cloud.

In the second type of hybridization, the authors of [50] proposed a hybrid algorithm called SFLA-GA which combines basic GA with SFLA (shuffled frog leaping algorithm) for seeking the ideal allocation of resources in CC. Because of the easier implementation and faster convergence of SFLA, this algorithm was widely used in the area of CC with the algorithm of CS for allocating the optimal resources [24]. Recently, Torabi and Safi-Esfahani proposed a solution for independent jobs scheduling in dynamic environment using a hybrid algorithm known as IRRO-CSO [92]. Their method tried to combine improved RRO (Raven Roosting Optimization) with CSO (Chicken Swarm Optimization) with the purpose of reducing the time of response and execution and boosting the rate of throughput computing. In this paper, CSO was employed for ensuring the harmony among the global and local searches, whereas the IRRO was used for finding a global solution.

1.3.10   Multi-Objective Swarm Optimization

Recently, some works are proposed as multi-objective optimization problems for solving workflow and job scheduling in CC. Reddy and Kumar in 2017 proposed a method for scheduling tasks based on multi-objective WOA [79]. In their study, the quality function was controlled by three factors known as energy, the use of machine, and the QoS. The performance of MOWOA was quantified by the minimum value of fitness. The simulation was realized using the CloudSim tool with JAYA. The results of the simulation present a superior performance compared with other metaheuristics algorithms such as ACO and PSO in terms of the time, the availability of resources, the scheduling cost, and energy.

In Ref. [44], Jena aimed to schedule jobs using a version of multi-objective PSO in CC called Nested PSO. Their study was focused on task scheduling by optimizing the processing time and energy. Jena used CloudSim platform for simulation and the obtained results showed superior quality according to multi-criteria. In Ref. [49], the authors proposed an algorithm for multi-objective scheduling based on multi-objective bacteria foraging optimization algorithm (MOBFOA). Their study consists of schedule independent jobs by minimizing flow time, makespan, and the cost of resource utilization. The proposed method showed a good performance compared with NSGA-II and optimized MOPSO in terms of measured criteria.

Another study for task scheduling in CC was developed recently by Srichandan et al. in 2018, based on hybrid MOBFOA in heterogeneous cloud environment [87]. The authors aimed to reduce both, the criteria of makespan and consumption of energy, simultaneously. The hybrid algorithm employed the operators of exploration/exploitation used in the GA to BFA with the purpose to enhance the convergence. The obtained results depicted that the hybrid MOBFOA provided high quality in contrast to other approaches according to the convergence, stability, and the diversity of the population.

In Ref. [88], the authors proposed NSGA-II based multi-objective job scheduling for CC. Their work aimed to decrease the use of energy and to minimize the makespan. Furthermore, the neural network was used for predicting the accessibility of the resource in a cloud. The numerical results of single objective based on makespan and consumption energy showed that the GA decreased effectively the energy consumption by incorporating the Dynamic Voltage Frequency Scaling (DVFS) in optimization procedure, however, the makespan increased. In addition, the multi-objective NSGA-II showed that the optimization procedure boosted by neural network provided a better solution. In Ref. [77], the authors considered four conflicting measures, known as minimizing job transfer time, job execution cost, energy consumption, and job queue length for devoting task based two multi-objective evolutionary algorithms called MOPSO and MOGA. From their study, the authors concluded that MOPSO was rapid and more efficient than MOGA for dealing job scheduling in CC.

Gabi et al. [37] proposed a multi-objective approach (MOA) based on a combination between the cat swarm algorithm and SA known as CSM-CSOSA. They used the OT technique with the purpose of enhancing the SA and providing better performance. In addition, they used a MO-QoS model for evaluating the performance of the CSM-CSOSA by CloudSim under two datasets. Quantitative analysis of CSM-CSOSA was evaluated with measures of cost, QoS, time, and performance enhancement rate. Meantime, the scalability analysis of CSM-CSOSA utilizing Isospeed-efficiency scalability metric was evenly defined. The experimental study showed that the CSM-CSOSA outperformed MOGA, MOACO, and MOPSO by providing lower values of time and cost. The presented research in [45] was focused on job scheduling for CC using multi-objective ABC (MO-ABC). Their objective was to reduce the processing time, energy, and cost resource. The proposed algorithm (TA-ABC) was tested by CloudSim and a comparative study was realized between TA-ABC and other algorithms. The experimental study showed that the MO-ABC reduced energy consumption and time execution by 32% and 26%, respectively in comparison with the MASA and the RSA.

In Ref. [100], the authors proposed a multi-swarm optimization algorithm (MSMOOA) for parallel dependent workflow scheduling. Their work aims to find a compromise between makespan, cost, and energy consumption. Two classes of particles in each swarm are generated. The first class communicates with several swarms at the same time, whereas the second class can only permutate the information within particles in the same swarm. In Ref. [104], the authors presented a MOACO for task scheduling. The authors considered two constraints of cost and deadline in CC. The experimental results showed a high performance in terms of makespan, the cardinal of a deadline, cost, and the rate of private machine employment. Chandrasekaran and Simon [14] proposed a hybridization method between CS algorithm and fuzzy system for solving MO-unit commitment problem. For solving such problem, two versions of CSA were generated—binary version of CSA and real version of CSA. The real version of CS can deal with the economic dispatch problem. The proposed technique was evaluated for single and MO-problem. The superiority of the proposed algorithm was identified on several unit test systems through the comparison of it with some other recent techniques via performance.

Bilgaiyan et al. [20] applied cat swarm-based MOOA for workflows scheduling in CC environment. Their work aimed to decrease the CPU time, cost, and makespan. The performance of the proposed algorithm was proved by a comparison with MOPSO technique. The author proved that the swarm-based solutions for such problems were the best suitable due to the requirements of users. So, MOOA became more used for workflow scheduling.

1.4   Conclusion

Task/Job scheduling in CC is a popular issue and plays an essential role in improving reliable and flexible platforms. Therefore, the researchers considered it as an optimization problem. Jobs scheduling works to map the resources to jobs, whereas the optimal scheduling method should guarantee a QoS and good performance of the CC by considering some criteria such as computation time and cost. The good results of swarm intelligence methods in solving many optimization problems led for their application in this field. Various swarm intelligence methods are applied for solving task scheduling issues in CC platforms. In this chapter, an overview of swarm intelligence for solving the problem of task scheduling in CC was presented including PSO, ABC, LOA, WAO, Bat algorithm, GWO, CSA, cat optimization algorithm, hybrid swarm algorithms, and multi-objective swarm optimization. The experimental results of the previous studies showed their ability to solve different types of scheduling tasks in CC. Most of these studies used a simulated environment to perform their experiments. The most used environments were CloudSim and CloudGrid. In addition, the most used number of VM varied from 5 to 2000 and the tasks ranged from 10 to 4000. Several studies solved this optimization problem using one swarm intelligence method, whereas the others used two or more methods to solve this problem as well as some studies treated with task scheduling as a multi-objective problem because they tried to minimize many factors along with makespan value. All the previous studies did not reach the optimum results and the field is still wide to perform more and more experiments in the future to enhance and improve the services of CC.

Bibliography

[1] Mohamed Abdel-Basset, Gunasekaran Manogaran, Doaa El-Shahat, and Seyedali Mirjalili. A hybrid whale optimization algorithm based on local search strategy for the permutation flow shop scheduling problem. *Future Generation Computer Systems*, 85:129–145, 2018.

[2] Mohit Agarwal and Gur Mauj Saran Srivastava. A cuckoo search algorithm-based task scheduling in cloud computing. In *Advances in Computer and Computational Sciences*, Singapore, pages 293–299. Springer, 2018.

[3] Pablo Eliseo Reynoso Aguirre, Flores Pérez Pedro, and María de Guadalupe Cota Ortiz. Multi-objective optimization using bat algorithm to solve multiprocessor scheduling and workload allocation problem. *Computer Science*, 2(2):41–51, 2015.

[4] Khaled Ahmed, Ahmed A Ewees, Mohamed Abd El Aziz, Aboul Ella Hassanien, Tarek Gaber, Pei-Wei Tsai, and Jeng-Shyang Pan. A hybrid krill-anfis model for wind speed forecasting. In *International Conference on Advanced Intelligent Systems and Informatics*, Cham, pages 365–372. Springer, 2016.

[5] Khaled Ahmed, Ahmed A Ewees, and Aboul Ella Hassanien. Prediction and management system for forest fires based on hybrid flower pollination optimization algorithm and adaptive neuro-fuzzy inference system. In *2017 Eighth International Conference on Intelligent Computing and Information Systems (ICICIS)*, Cairo, pages 299–304. IEEE, 2017.

[6] Mohammed AA Al-qaness, Mohamed Abd Elaziz, and Ahmed A Ewees. Oil consumption forecasting using optimized adaptive neuro-fuzzy inference system based on sine cosine algorithm. *IEEE Access*, 6:68394–68402, 2018.

[7] Entisar S Alkayal, Nicholas R Jennings, and Maysoon F Abulkhair. Efficient task scheduling multi-objective particle swarm optimization in cloud computing. In *2016 IEEE 41st Conference on Local Computer Networks Workshops (LCN Workshops)*, Dubai, pages 17–24. IEEE, 2016.

[8] Nora Almezeini and Alaaeldin Hafez. Task scheduling in cloud computing using lion optimization algorithm. *Algorithms*, 5:7, 2017.

[9] Supacheep Amtade and Toshiyuki Miyamoto. Cuckoo search algorithm for job scheduling in cloud systems. *IEICE Transactions on Fundamentals of Electronics, Communications and Computer Sciences*, 98(2):645–649, 2015.

[10] KR Remesh Babu and Philip Samuel. Enhanced bee colony algorithm for efficient load balancing and scheduling in cloud. In *Innovations in Bio-Inspired Computing and Applications*, Cham, pages 67–78. Springer, 2016.

[11] RG Babukartik and P Dhavachelvan. Hybrid algorithm using the advantage of aco and cuckoo search for job scheduling. *International Journal of Information Technology Convergence and Services*, 2(4):25, 2012.

[12] Balamurugan Balusamy, Jayashree Sridhar, Divya Dhamodaran, and P Venkata Krishna. Bio-inspired algorithms for cloud computing: a review. *International Journal of Innovative Computing and Applications*, 6(3–4):181–202, 2015.

[13] AS Ajeena Beegom and MS Rajasree. A particle swarm optimization based pareto optimal task scheduling in cloud computing. In *International Conference in Swarm Intelligence*, Cham, pages 79–86. Springer, 2014.

[14] Ashish Kumar Bhandari, Vineet Kumar Singh, Anil Kumar, and Girish Kumar Singh. Cuckoo search algorithm and wind driven optimization based study of satellite image segmentation for multilevel thresholding using kapur's entropy. *Expert Systems with Applications*, 41(7):3538–3560, 2014.

[15] Saurabh Bilgaiyan, Santwana Sagnika, and Madhabananda Das. Workflow scheduling in cloud computing environment using cat swarm optimization. In *2014 IEEE International conference on Advance Computing Conference (IACC)*, Gurgaon, pages 680–685. IEEE, 2014.

[16] Saurabh Bilgaiyan, Santwana Sagnika, and Madhabananda Das. A multi-objective cat swarm optimization algorithm for workflow scheduling in cloud computing environment. In *Intelligent Computing, Communication and Devices*, New Delhi, pages 73–84. Springer, 2015.

[17] Shashikant Burnwal and Sankha Deb. Scheduling optimization of flexible manufacturing system using cuckoo search-based approach. *The International Journal of Advanced Manufacturing Technology*, 64(5–8):951–959, 2013.

[18] K Chandrasekaran and Sishaj P Simon. Multi-objective scheduling problem: hybrid approach using fuzzy assisted cuckoo search algorithm. *Swarm and Evolutionary Computation*, 5:1–16, 2012.

[19] Amlan Chatterjee, Matthew Levan, Crosby Lanham, and Mishael Zerrudo. Job scheduling in cloud datacenters using enhanced particle swarm optimization. In *2017 2nd International Conference for Convergence in Technology (I2CT)*, Mumbai, pages 895–900. IEEE, 2017.

[20] Prachi Chaturvedi, Abhishek Satyarthi, and Sanjiv Sharma. Time and reliability optimization bat algorithm for scheduling workflow in cloud. *International Research Journal of Engineering and Technology*, 4(6), 2017.

[21] Huangning Chen and Wenzhong Guo. Real-time task scheduling algorithm for cloud computing based on particle swarm optimization. In *International Conference on Cloud Computing and Big Data in Asia*, Cham, pages 141–152. Springer, 2015.

[22] Shu-Chuan Chu and Pei-Wei Tsai. Computational intelligence based on the behavior of cats. *International Journal of Innovative Computing, Information and Control*, 3(1):163–173, 2007.

[23] J Devagnanam and NM Elango. Design and development of exponential lion algorithm for optimal allocation of cluster resources in cloud. *Cluster Computing*, 22:1385–1400, 2018.

[24] P Durgadevi and S Srinivasan. Resource allocation in cloud computing using SFLA and cuckoo search hybridization. *International Journal of Parallel Programming*, 112:1–17, 2018.

[25] Russell Eberhart and James Kennedy. A new optimizer using particle swarm theory. In *Proceedings of the Sixth International Symposium on Micro Machine and Human Science, 1995 (MHS'95)*, Nagoya, pages 39–43. IEEE, 1995.

[26] Mohamed Abd El Aziz, Ahmed A Ewees, and Aboul Ella Hassanien. Hybrid swarms optimization based image segmentation. In *Hybrid Soft Computing for Image Segmentation*, Cham, pages 1–21. Springer, 2016.

[27] Mohamed Abd El Aziz, Ahmed A Ewees, and Aboul Ella Hassanien. Whale optimization algorithm and moth-flame optimization for multilevel thresholding image segmentation. *Expert Systems with Applications*, 83:242–256, 2017.

[28] Mohamed E Abd Elaziz, Ahmed A Ewees, Diego Oliva, Pengfei Duan, and Shengwu Xiong. A hybrid method of sine cosine algorithm and differential evolution for feature selection. In *International Conference on Neural Information Processing*, Cham, pages 145–155. Springer, 2017.

[29] Gamal F Elhady and Medhat A Tawfeek. A comparative study into swarm intelligence algorithms for dynamic tasks scheduling in cloud computing. In *2015 IEEE Seventh International Conference on Intelligent Computing and Information Systems (ICICIS)*, Cairo, pages 362–369. IEEE, 2015.

[30] Ahmed A Ewees, Mohamed Abd El Aziz, and Mohamed Elhoseny. Social-spider optimization algorithm for improving Anfis to predict biochar yield. In *2017 8th International Conference on Computing, Communication and Networking Technologies (ICCCNT)*, Delhi, pages 1–6. IEEE, 2017.

[31] Ahmed A Ewees, Mohamed Abd El Aziz, and Aboul Ella Hassanien. Chaotic multi-verse optimizer-based feature selection. *Neural Computing and Applications*, 31:1–16, 2017.

[32] Ahmed A Ewees and Mohamed Abd Elaziz. Improved adaptive neuro-fuzzy inference system using gray wolf optimization: a case study in predicting biochar yield. *Journal of Intelligent Systems*, 29(1):924–940, 2018.

[33] Ahmed A Ewees, Mohamed Abd Elaziz, and Diego Oliva. Image segmentation via multilevel thresholding using hybrid optimization algorithms. *Journal of Electronic Imaging*, 27(6):063008, 2018.

[34] Ahmed A Ewees and Ahmed T Sahlol. Bio-inspired optimization algorithms for improving artificial neural networks: a case study on handwritten letter recognition. *Computational Linguistics, Speech and Image Processing for Arabic Language*, 4:249, 2018.

[35] Fereshteh Ershad Farkar and Ali Asghar Pourhaji Kazem. Bi-objective task scheduling in cloud computing using chaotic bat algorithm. *International Journal of Advanced Computer Science and Applications*, 8(10):223–230, 2017.

[36] Danlami Gabi, Abdul Samad Ismail, Anazida Zainal, Zalmiyah Zakaria, and Ajith Abraham. Orthogonal taguchi-based cat algorithm for solving task scheduling problem in cloud computing. *Neural Computing and Applications*, 30(6):1845–1863, 2018.

[37] Danlami Gabi, Abdul Samad Ismail, Anazida Zainal, Zalmiyah Zakaria, and Ahmad Al-Khasawneh. Hybrid cat swarm optimization and simulated annealing for dynamic task scheduling on cloud computing environment. *Journal of ICT*, 17(3):435–467, 2018.

[38] Amir Hossein Gandomi, Xin-She Yang, Amir Hossein Alavi, and Siamak Talatahari. Bat algorithm for constrained optimization tasks. *Neural Computing and Applications*, 22(6):1239–1255, 2013.

[39] Samer Hanoun, Saeid Nahavandi, Doug Creighton, and Hans Kull. Solving a multiobjective job shop scheduling problem using pareto archived cuckoo search. In *2012 IEEE 17th Conference on Emerging Technologies & Factory Automation (ETFA)*, Krakow, pages 1–8. IEEE, 2012.

[40] Rehab Ali Ibrahim, Mohamed Abd Elaziz, Ahmed A Ewees, Ibrahim M Selim, and Songfeng Lu. Galaxy images classification using hybrid brain storm optimization with moth flame optimization. *Journal of Astronomical Telescopes, Instruments, and Systems*, 4(3):038001, 2018.

[41] S Induja and VP Eswaramurthy. Bat algorithm: an overview and its applications. *International Journal of Advanced Research in Computer and Communication Engineering*, 5(1):448–451, 2016.

[42] Liji Jacob. Bat algorithm for resource scheduling in cloud computing. *International Journal of Engineering Sciences and Research Technology*, 2:53–57, 2014.

[43] Bappaditya Jana, Moumita Chakraborty, and Tamoghna Mandal. A task scheduling technique based on particle swarm optimization algorithm in cloud environment. In *Soft Computing: Theories and Applications*, Singapore, pages 525–536. Springer, 2019.

[44] RK Jena. Multi objective task scheduling in cloud environment using nested PSO framework. *Procedia Computer Science*, 57:1219–1227, 2015.

[45] RK Jena. Task scheduling in cloud environment: a multi-objective ABC framework. *Journal of Information and Optimization Sciences*, 38(1):1–19, 2017.

[46] Mala Kalra and Sarbjeet Singh. A review of metaheuristic scheduling techniques in cloud computing. *Egyptian Informatics Journal*, 16(3):275–295, 2015.

[47] G Kanagaraj, SG Ponnambalam, N Jawahar, and J Mukund Nilakantan. An effective hybrid cuckoo search and genetic algorithm for constrained engineering design optimization. *Engineering Optimization*, 46(10):1331–1351, 2014.

[48] Dervis Karaboga. An idea based on honey bee swarm for numerical optimization. Technical report, Technical report-tr06, Erciyes University, Engineering Faculty, Computer Engineering Department, Turkey, 2005.

[49] Mandeep Kaur and Sanjay Kadam. A novel multi-objective bacteria foraging optimization algorithm (MOBFOA) for multi-objective scheduling. *Applied Soft Computing*, 66:183–195, 2018.

[50] S Kayalvili and M Selvam. Hybrid SFLA-GA algorithm for an optimal resource allocation in cloud. *Cluster Computing*, 22(2):3165–3173, 2019.

[51] Azade Khalili and Seyed Morteza Babamir. Optimal scheduling work-flows in cloud computing environment using pareto-based grey wolf optimizer. *Concurrency and Computation: Practice and Experience*, 29(11):e4044, 2017.

[52] P Venkata Krishna. Honey bee behavior inspired load balancing of tasks in cloud computing environments. *Applied Soft Computing*, 13(5):2292–2303, 2013.

[53] Sunil Kumar and Mala Kalra. A hybrid approach for energy-efficient task scheduling in cloud. In *Proceedings of 2nd International Conference on Communication, Computing and Networking*, Singapore, pages 1011–1019. Springer, 2019.

[54] V Suresh Kumar and M Aramudhan. Trust based resource selection in cloud computing using hybrid algorithm. *International Journal of Intelligent Systems and Applications*, 7(8):59, 2015.

[55] V Suresh Kumar et al. Hybrid optimized list scheduling and trust based resource selection in cloud computing. *Journal of Theoretical & Applied Information Technology*, 69(3), 2014.

[56] Xiangtao Li and Minghao Yin. A hybrid cuckoo search via lévy flights for the permutation flow shop scheduling problem. *International Journal of Production Research*, 51(16):4732–4754, 2013.

[57] Fei Luo, Ye Yuan, Weichao Ding, and Haifeng Lu. An improved par-ticle swarm optimization algorithm based on adaptive weight for task scheduling in cloud computing. In *Proceedings of the 2nd International Conference on Computer Science and Application Engineering*, Hohhot, page 142. ACM, 2018.

[58] Rakesh Madivi and S Sowmya Kamath. An hybrid bio-inspired task scheduling algorithm in cloud environment. In *2014 International Conference on Computing, Communication and Networking Technologies (ICCCNT)*, pages 1–7. IEEE, 2014.

[59] Syed Hamid Hussain Madni, Muhammad Shafie Abd Latiff, Yahaya Coulibaly, Resource scheduling for infrastructure as a service (IAAS) in cloud computing: challenges and opportunities. *Journal of Network and Computer Applications*, 68:173–200, 2016.

[60] Behnam Malakooti, Shaya Sheikh, Camelia Al-Najjar, and Hyun Kim. Multi-objective energy aware multiprocessor scheduling using bat intel-ligence. *Journal of Intelligent Manufacturing*, 24(4):805–819, 2013.

[61] Tripti Mandal and Sriyankar Acharyya. Optimal task scheduling in cloud computing environment: meta heuristic approaches. In *2015 2nd International Conference on Electrical Information and Communication Technology (EICT)*, pages 24–28. IEEE, 2015.

[62] MK Marichelvam, T Prabaharan, Xin-She Yang, and M Geetha. Solving hybrid flow shop scheduling problems using bat algorithm. *International Journal of Logistics Economics and Globalisation*, 5(1):15–29, 2013.

[63] MK Marichelvam, Thirumoorthy Prabaharan, and Xin-She Yang. Improved cuckoo search algorithm for hybrid flow shop scheduling problems to minimize makespan. *Applied Soft Computing*, 19:93–101, 2014.

[64] Mohand Mezmaz, Nouredine Melab, Yacine Kessaci, Young Choon Lee, E-G Talbi, Albert Y Zomaya, and Daniel Tuyttens. A parallel bi-objective hybrid metaheuristic for energy-aware scheduling for cloud computing systems. *Journal of Parallel and Distributed Computing*, 71(11):1497–1508, 2011.

[65] Seyedali Mirjalili and Andrew Lewis. The whale optimization algorithm. *Advances in Engineering Software*, 95:51–67, 2016.

[66] Seyedali Mirjalili, Seyed Mohammad Mirjalili, and Andrew Lewis. Grey wolf optimizer. *Advances in Engineering Software*, 69:46–61, 2014.

[67] Gobalakrishnan Natesan and Arun Chokkalingam. Opposition learning-based grey wolf optimizer algorithm for parallel machine scheduling in cloud environment. *International Journal of Intelligent Engineering and Systems*, 10(1):186–195, 2017.

[68] Gobalakrishnan Natesan and Arun Chokkalingam. Task scheduling in heterogeneous cloud environment using mean grey wolf optimization algorithm. *ICT Express*, 5(2):110–114, 2019.

[69] Nima Jafari Navimipour and Farnaz Sharifi Milani. Task scheduling in the cloud computing based on the cuckoo search algorithm. *International Journal of Modeling and Optimization*, 5(1):44, 2015.

[70] Aziz Ouaarab, Belaïd Ahiod, Xin-She Yang, and Mohammed Abbad. Discrete cuckoo search algorithm for job shop scheduling problem. In *2014 IEEE International Symposium on Intelligent Control (ISIC)*, Juan Les Pins, pages 1872–1876. IEEE, 2014.

[71] Jeng-Shyang Pan, Haibin Wang, Hongnan Zhao, and Linlin Tang. Interaction artificial bee colony based load balance method in cloud computing. In *Genetic and Evolutionary Computing*, Cham, pages 49–57. Springer, 2015.

[72] S Raghavan, P Sarwesh, C Marimuthu, and K Chandrasekaran. Bat algorithm for scheduling workflow applications in cloud. In *2015 International Conference on Electronic Design, Computer Networks & Automated Verification (EDCAV)*, Shillong, pages 139–144. IEEE, 2015.

[73] BR Rajakumar. Lion algorithm for standard and large scale bilinear system identification: a global optimization based on lion's social behavior. In *2014 IEEE Congress on Evolutionary Computation (CEC)*, pages 2116–2123. IEEE, 2014.

[74] R Raju, RG Babukarthik, D Chandramohan, P Dhavachelvan, and T Vengattaraman. Minimizing the makespan using hybrid algorithm for cloud computing. In *Advance Computing Conference (IACC), 2013 IEEE 3rd International*, Ghaziabad, pages 957–962. IEEE, 2013.

[75] Fahimeh Ramezani, Jie Lu, and Farookh Hussain. Task scheduling optimization in cloud computing applying multi-objective particle swarm optimization. In *International Conference on Service-Oriented Computing*, pages 237–251. Springer, 2013.

[76] Fahimeh Ramezani, Jie Lu, and Farookh Khadeer Hussain. Task-based system load balancing in cloud computing using particle swarm optimization. *International Journal of Parallel Programming*, 42(5):739–754, 2014.

[77] Fahimeh Ramezani, Jie Lu, Javid Taheri, and Farookh Khadeer Hussain. Evolutionary algorithm-based multi-objective task scheduling optimization model in cloud environments. *World Wide Web*, 18(6):1737–1757, 2015.

[78] T Sunitha Rani and Shyamala Kannan. Task scheduling on virtual machines using bat strategy for efficient utilization of resources in cloud environment. *International Journal of Applied Engineering Research*, 12(17):6663–6669, 2017.

[79] G Narendrababu Reddy and S Phani Kumar. Multi objective task scheduling algorithm for cloud computing using whale optimization technique. In *International Conference on Next Generation Computing Technologies*, Singapore, pages 286–297. Springer, 2017.

[80] Santwana Sagnika, Saurabh Bilgaiyan, and Bhabani Shankar Prasad Mishra. Workflow scheduling in cloud computing environment using bat algorithm. In *Proceedings of First International Conference on Smart System, Innovations and Computing*, Singapore, pages 149–163. Springer, 2018.

[81] Ahmed T Sahlol, Ahmed A Ewees, Ahmed Monem Hemdan, and Aboul Ella Hassanien. Training feedforward neural networks using

sine-cosine algorithm to improve the prediction of liver enzymes on fish farmed on nano-selenite. In *2016 12th International Computer Engineering Conference (ICENCO)*, Cairo, pages 35–40. IEEE, 2016.

[82] MR Sathya and M Mohamed Thameem Ansari. Load frequency control using bat inspired algorithm based dual mode gain scheduling of pi controllers for interconnected power system. *International Journal of Electrical Power & Energy Systems*, 64:365–374, 2015.

[83] S Selvarani and G Sudha Sadhasivam. Improved cost-based algorithm for task scheduling in cloud computing. In *2010 IEEE International Conference on Computational Intelligence and Computing Research (ICCIC)*, pages 1–5. IEEE, 2010.

[84] Shabnam Sharma, Ashish Kr Luhach, and Kiran Jyoti. A novel approach of load balancing in cloud computing using computational intelligence. *International Journal of Engineering and Technology*, 8(1):124–128, 2016.

[85] YS Sheeja and S Jayalekshmi. Cost effective load balancing based on honey bee behaviour in cloud environment. In *2014 First International Conference on Computational Systems and Communications (ICCSC)*, Trivandrum, pages 214–219. IEEE, 2014.

[86] Poonam Singh, Maitreyee Dutta, and Naveen Aggarwal. A review of task scheduling based on meta-heuristics approach in cloud computing. *Knowledge and Information Systems*, 52(1):1–51, 2017.

[87] Srichandan Sobhanayak, Ashok Kumar Turuk, and Bibhudatta Sahoo. Task scheduling for cloud computing using multi-objective hybrid bacteria foraging algorithm. *Future Computing and Informatics Journal*, 3:210–230, 2018.

[88] A Sathya Sofia and P GaneshKumar. Multi-objective task scheduling to minimize energy consumption and makespan of cloud computing using nsga-ii. *Journal of Network and Systems Management*, 26(2):463–485, 2018.

[89] Karnam Sreenu and M Sreelatha. W-scheduler: whale optimization for task scheduling in cloud computing. *Cluster Computing*, 22:1087–1098, 2019.

[90] Mahalaxmi Sridhar and G Rama Mohan Babu. Hybrid particle swarm optimization scheduling for cloud computing. In *2015 IEEE International Advance Computing Conference (IACC)*, Banglore, India, pages 1196–1200. IEEE, 2015.

[91] Avnish Thakur and Major Singh Goraya. A taxonomic survey on load balancing in cloud. *Journal of Network and Computer Applications*, 98:43–57, 2017.

[92] Shadi Torabi and Faramarz Safi-Esfahani. A dynamic task scheduling framework based on chicken swarm and improved raven roosting optimization methods in cloud computing. *The Journal of Supercomputing*, 74(6):2581–2626, 2018.

[93] R Valarmathi and T Sheela. Ranging and tuning based particle swarm optimization with bat algorithm for task scheduling in cloud computing. *Cluster Computing*, 22(5):11975–11988, 2019.

[94] Hui Wang, Wenjun Wang, Hui Sun, Zhihua Cui, Shahryar Rahnamayan, and Sanyou Zeng. A new cuckoo search algorithm with hybrid strategies for flow shop scheduling problems. *Soft Computing*, 21(15):4297–4307, 2017.

[95] Zhiming Wang, Kai Shuang, Long Yang, and Fangchun Yang. Energy-aware and revenue-enhancing combinatorial scheduling in virtualized of cloud datacenter. *Journal of Convergence Information Technology*, 7(1):62–70, 2012.

[96] William A Watkins and William E Schevill. Aerial observation of feeding behavior in four baleen whales: Eubalaena glacialis, balaenoptera borealis, megaptera novaeangliae, and balaenoptera physalus. *Journal of Mammalogy*, 60(1):155–163, 1979.

[97] Daqin Wu. Cloud computing task scheduling policy based on improved particle swarm optimization. In *2018 International Conference on Virtual Reality and Intelligent Systems (ICVRIS)*, Changsha, China, pages 99–101. IEEE, 2018.

[98] Xin-She Yang. A new metaheuristic bat-inspired algorithm. In *Nature Inspired Cooperative Strategies for Optimization (NICSO 2010)*, Berlin, Heidelberg, pages 65–74. Springer, 2010.

[99] Xin-She Yang and Suash Deb. Cuckoo search via lévy flights. In *World Congress on Nature & Biologically Inspired Computing, 2009. NaBIC 2009*, Coimbatore, India, pages 210–214. IEEE, 2009.

[100] Guang-shun Yao, Yong-sheng Ding, and Kuang-rong Hao. Multi-objective workflow scheduling in cloud system based on cooperative multi-swarm optimization algorithm. *Journal of Central South University*, 24(5): 1050–1062, 2017.

[101] Maziar Yazdani and Fariborz Jolai. Lion optimization algorithm (loa): a nature-inspired metaheuristic algorithm. *Journal of Computational Design and Engineering*, 3(1):24–36, 2016.

[102] Selim Yılmaz and Ecir U Küçüksille. A new modification approach on bat algorithm for solving optimization problems. *Applied Soft Computing*, 28:259–275, 2015.

[103] Fuqing Zhao and Jianxin Tang. A memetic algorithm combined particle swarm optimization with simulated annealing and its application on multiprocessor scheduling problem. *PRZ Elektrotechniczny*, 88:292–296, 2012.

[104] Liyun Zuo, Lei Shu, Shoubin Dong, Yuanfang Chen, and Li Yan. A multi-objective hybrid cloud resource scheduling method based on deadline and cost constraints. *IEEE Access*, 5:22067–22080, 2017.

2

Techniques for Resource Sharing in Cloud Computing Platform

B.L. Radhakrishnan, S. Sudhakar, R.V. Belfin, P. Karthikeyan, E. Kirubakaran, and K. Martin Sagayam

Karunya Institute of Technology and Sciences

CONTENTS

2.1   Introduction

Resource sharing is a way to share the resources with other users or providers in the cloud. Cost of IT involves capital expenditure, operational expenditure, and total cost of ownership. This cost of resources is a barrier to the small- and medium-sized service providers or customers. Also, the underutilization of the resources brings down the profit of the provider or customer. The service providers and customers share their resources to reduce the cost and increase the resource utilization. The small- and medium-size providers are unable to meet the dynamic resource requirement due to lack of resources. Multiple providers or customers aggregate their unused resources to create a resource pool. The dynamic demand of IT can be meet out easily by the resource pool. The resource sharing at different clouds is depicted in Figure 2.1.

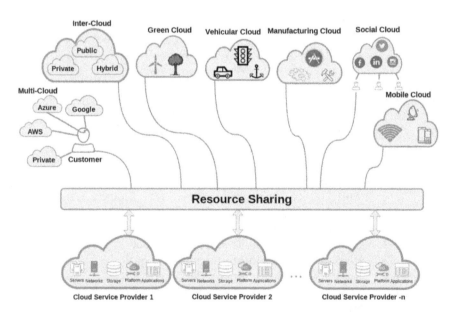

FIGURE 2.1
Resource sharing at different clouds.

Inter Cloud (IC) integrates many cloud providers to exchange resources among themselves. IC operates in two ways: (1) cloud Federation (CF) and (2) Multi-Cloud (MC). CF is a group of service providers that voluntarily share their resources. MC customers use different service provider's resources to host their applications. Social media platform enables people to collaborate for resource sharing. Like-minded people create groups in social media to share, rent, and barter their resources that forms Social Cloud (SC). So far smartphone market has attracted more than 2.5 billion people worldwide. Smartphone has resource limitations, which is overcome by collaboration among smart devices and Cloud. Mobile cloud extends the mobile computing by using cloud providers, and wireless network operators to enrich the functionality.

The Small and Medium Enterprises (SMEs) in manufacturing have resource and capacity shortage. Sharing resources increases productivity and reduces the cost of production. Manufacturing Cloud (MC) simplifies the resource sharing among SMEs. Resource sharing in MC happens in two environments: (1) IT and (2) Manufacturing. Modern vehicles possess reasonable computing devices. Pooling of computing resources increases the computational capability of vehicles. The resource pool is used to perform the computation of traffic management, multimedia services, and smart parking. Vehicular Cloud (VC) enables the connected vehicles to process and share information with others. Energy saving is significant in today's world. Energy-aware resource utilization reduces energy consumption and pollution. supports energy-aware resource sharing, which saves energy, reduces carbon emission and water consumption. The following sections describe various resource sharing methods in clouds.

The remaining of the chapter is organized as follows: Section 2.2 discusses IC, Section 2.3 describes SC, Section 2.4 explains mobile cloud, Section 2.5 discusses manufacturing cloud, Section 2.6 describes VC, and Section 2.7 explains green cloud.

2.2 Intercloud

Interconnects clouds together to augment its capacity and resource utilization. It enables cloud providers to share underutilized resources to maximize profit. Based on the party collaboration, IC is categorized as volunteer federation (federated cloud) and independent cloud (multi-cloud) [1]. In a federated cloud, providers join together to share and exchange their resources. Federated cloud is subdivided into centralized cloud (agent-based resource allocation) and cloud (peer allocates resources). In a multi-cloud, providers are not needed to join together to exchange resources and customers are responsible for gathering and controlling resources from multiple providers. Services and libraries are the further classifications of multi-cloud. Figure 2.2 shows the classification of IC.

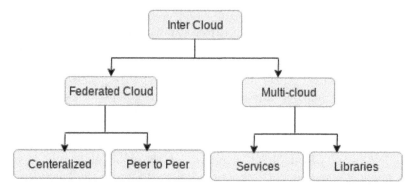

FIGURE 2.2
Classification of inter cloud.

2.2.1   Resource Sharing in Federated Cloud

2.2.1.1   Agent-Based (Centralized) Resource Sharing in Federated Cloud

An agent is an autonomous software system that discovers, matches, selects, negotiates, schedules, and monitors resources in the cloud [66]. This section discusses various approaches to agent-based resource sharing in the federated cloud environment.

A Multi-Agent System for Cloud Resource Enhancement: this Cloud Resource Bartering System (CRBS) implemented by Goher et al. is a platform for IaaS service provider to barter underutilized pooled resources without financial transaction [82]. Agents in CRBS enable resource exchange among multiple providers and handle the dynamic demand. Customer submits a resource request along with the following parameters to the user agent: resource type, number of resources required, maximum cost, resource urgency, and resource region. Multi-criteria based resource selection algorithm is used to handle the submitted request. The resource selection process considers the following factors, such as resource type, number of resources required, exchange time, maximum cost, resource urgency, and provider's rank while selecting resources. Bartering agent uses a price determination algorithm to satisfy both the customer and the provider during price negotiation. Finally, the control of the selected resources is transferred to the customer after price negotiation and service level agreement.

Automated Bartering System: this multi-agent based e-barter system developed by Demirkol et al. is an online system for IaaS service provider to exchange resources between customer and provider [22]. This e-barter system uses many intelligent software agents for trade management and resource matching. The resource matching in system uses ontology-based comparison. It uses an intelligent Semantic Web Service (SWS) agent to deduce closeness between the quoted price and the selling price. The customer agent adds resource request by sending a request for proposal message to the system. Barter agent discovers matching resource for the request and sends matched resource details to the customer agent. Then, negotiation between customer and provider is initiated. Customer agent sends, accept, or reject message to the system after the negotiation. Cargo agent is responsible for shipping resource control to the customer after a successful negotiation.

For Resource Matching: a Belief-Desire-Intension (BDI) agent-based semantic e-barter system proposed by Cakmaz et al. is a multi-agent system for IaaS Service provider to share resources [12]. uses a JACK framework to build agents in the system. When the customer submits a request, the BDI uses two ways to match resources—(1) exact matching: matching of resources as per request submitted by the customer and (2) semantic matching: matching of resources which are close to the request. First, the matchmaker agent applies the exact matching technique. If an exact match fails, then it applies the semantic matching. Second, the customer agent executes resource exchange plan; if the plan type is "no negotiation", it executes a negotiation plan using three ratios: minimum ratio, actual ratio, and an inverse ratio. Minimum ratio is the minimum ratio between the cost of the customer to the resource and the quoted cost of the provider. The actual ratio is the ratio between the quoted cost of the customer to the resource and quoted cost of the provider. The inverse ratio is the ratio between the quoted cost of the provider and the quoted customer cost to the resource. User-agent checks the customer Requested Ratio (RR) with the New Actual Ratio (NRR) using the following steps:

step 1: NRR $\leq$ RR,

if true bartering successful resource swapping happens else go to step 2

step 2: upper-ratio $\leq$ NRR

if true then bartering unsuccessful and the process terminated,

else select a new offer and repeat step 1.

The Multi-Agent System for Resource Selection: cloud service selection using an adaptive learning method developed by Wang et al. optimizes resource selection dynamically and gives a consolidated solution to the customer [74]. It executes many resource selection algorithms to select the best matching resources from the resource pool. The customer agent accepts the resource request, decomposes the submitted request into distinct categories of sub-request, and sends to an intermediate agent. Adaptive learning algorithms are used by the intermediate agent to perform resource selection and return selected resources to the customer agent for integration. Service providers register with federation through cloud service agent and submit the resource availability information. Cloud service agent monitors the resource performance and records performance after transferring.

2.2.1.2   Peer-to-Peer Resource Sharing in Federated Cloud

Enables peers to share resources among themselves without any centralized agent or third party. The following section describes various techniques used in peer-to-peer resource sharing in the federated cloud.

Resource Sharing in the Cross-Federated Cloud: cross-cloud federation (CCF) developed by Celesti et al. is a cloud testbed for horizontal federation [15]. CCF supports IaaS service providers to expand resource provisioning capabilities dynamically using Peer-to-Peer (P2P) approach. It consists of two clouds namely home cloud and foreign cloud. Home cloud makes resource request to the foreign cloud. The cloud manager in uses three agents such as discovery agent, matchmaking agent, and authentication agent for resource sharing. Discovery manager is responsible for preparing and publishing message across cloud providers. It uses "publish and subscribe" technique to acquire resources from other clouds. Match-making manager is responsible for selecting a correct resource, manage, and enforce policies of the cloud providers. The authentication manager creates, implements, and ensures security policies among cloud providers. It uses Single Sign On (SSO) technique to implement authentication in CCF.

Resource Sharing Using Incentive Technique: Cluster-Based Incentive Mechanism (CBIM) proposed by Zhang and Antonopoulos is an incentive-based resource sharing method [84]. CBIM creates resource providers cluster to share resources for servicing a resource request. The consumer has to initiate a transaction by sending a resource request query to all the providers in the cluster. The query consists of consumer id, address, resource request, and resource provision. The consumer has to share the same amount of other types of resource to other cloud providers in the cluster. uses the Eulerian graph forming technique to form a cluster. Once the cluster is formed, the resource sharing starts. The reputation value of the consumer is calculated based on the past resource sharing history. Each consumer is responsible for calculating the reputation of the other cloud providers in the ring before agreeing to share resources.

Resource Sharing Using Indirect Reciprocity: Fairness Driven Networks of Favor (FD-NoF) approach proposed by Falcao et al. enhances peer-to-peer cooperation while sharing resources with peers [26]. Cloud providers increase resource sharing capacity based on their resource sharing fairness value (using transaction history). A resource consumer in the group begins the transaction by sending a resource request to all the other members in the group. All the members of the group maintain transaction record of other members. The fairness value is obtained by considering the past resource contribution of the consumer. Satisfaction value is obtained by the amount of resource requests by the consumer and available resource with the provider. The "fairness value" and "satisfaction value" are used to process the resource request from the customer. Based on the calculated "fairness value" and "satisfaction value" providers share resources with other members. Finally, the transaction record is shared to the members of the group for maintaining history.

Multi-Tenant Based Resource Sharing: Reciprocal Resource Fairness (RRF) based resource sharing model proposed by Liu et al. enables fair resource sharing among multiple IaaS cloud customers [49]. RRF is a measure of the ratio between amounts of resources shared and amounts of resource consumed by the customer in a federation. Inter-tenant Resource Trading (IRT) is a resource trading technique which allows the exchange of resources among the customers. RRF uses IRT technique to share resources among customers. In RRF, initially, customers acquire the required resources from many cloud service providers. The acquired resources may be surplus or insufficient to meet the customer's actual requirement. Then, customers make a trade of surplus resource to meet an insufficient resource using IRT technique. RRF guarantees fairness in IRT trading (resource sharing). Intra-tenant Weight Adjustment (IWA) distributes workloads evenly to all the resources to avoid overloading a particular resource. Table 2.1 shows the comparison of centralized and peer-to-peer resource sharing techniques.

TABLE 2.1

Centralized System Versus Peer-to-Peer System

Federation Type	Techniques	Strengths	Challenges
Centralized	CRBS	Easier service discovery	Implementation complexity
	MAS based e-barter system	High security	Free riders
	Semantic e-barter system	Trust worthiness	Ensuring fairness
	Adaptive learning mechanism	Easier transaction maintenance	New comers
Peer to Peer	CCF	Quality of service incentive for resource sharing	Finding free resources
	CBIM	Avoids free riders	Maintaining transaction record
	FDNoF	Less implementation complexity	Security
	RRF	Motivates new comers	Trust worthiness
			Difficulties in ensuring quality

2.2.2   Resource Sharing in Multi-Cloud

2.2.2.1   Agent-Based Service Selection Approaches in Multi-Cloud

Self-coordinating multi-agent based service selection model proposed by Gutierrez-Garcia et al. collects services from multiple service providers and consolidates them as a single service [32]. It uses multiple agents to orchestrate, coordinate, and control services from many service providers. Service Composition Table (SCT) consists of information such as participants list, services list, services status, and agents list. Semi-recursive contract net protocol (SR-CNP) augmented with a service composition table technique is proposed in this system. Every agent has its own SCT table. The Customer Agent (CA) prepares service request using the service ontology and sends to the Brokering Agent (BA). BA refers the SCT table for service availability and sends the "request for proposal" message to the Service Provider Agents (SPAs). SPA checks the resource status (available, busy, and failed) in consultation with the Resource Agents (RAs) and sends the reply back to the BA. BA manages all the SPAs communications, makes services from multiple SPAs as a single service, and gives it to the customer.

Focused Selection Contract Net Protocol (FSCNP) proposed by Kwang Mong Sim gathers services from many service providers and makes as a single service [65]. FSCNP uses many agents to discover, find similarities, filter, and select services from many providers. Cloudle is a search engine based on MAS for finding services in the cloud. Customers request their services using Cloudle. The customer request consists of functional requirements, technical requirements, and cost of the services. The Service Discovery Agent (SDA) performs the similarity matching using the keyword taken from the request. Then, the matched service is selected by the SDA. Cloud crawler is an agent which accumulates services from many providers. SDA uses cloud crawler to aggregate service provider information. BA determines offered price using two techniques: (1) market-driven negotiation strategy, and (2) bargaining position gaining strategy. Finally, BA selects the service provider based on the best offer and sends the details to the customer.

Ontology-based service discovery: an ontology-based recommendation approach for matching service provider proposed by Kang et al. helps customers to find suitable service provider thorough intelligent brokering system [25]. Cloud ontology is a concept used to find similarities among cloud services. The customer submits a service request to the system using CA. BA uses cloud ontology to find similarities between requested services and available services. BAs use three types of similarities (concept similarity, object property similarity, and database type similarity) to select the most suitable service providers from the providers' list. Initially, BA selects services based on service availability. Secondly, it evaluates selected services using similarities relationship. Thirdly, it removes services which have low similarity value from the initially selected services. Finally, the selected service is sent to the customer, if BA selects services successfully, else BA send a service request to the other BAs which are capable of connecting to more service providers.

2.2.2.2   QoS-Based Service Selection in Multi-Cloud

Quality of service (QoS) and customer satisfaction level vary depending upon the hosting environment and availability zone. The above problem leads to uncertainty in QoS and user satisfaction levels. Kernel Density Estimation (KDE) based recommendation system developed by Haithem solved this uncertainty problem when services are hosted in multi-cloud [55]. KDE is an estimation method used to predict the probability of occurrence. Customer satisfaction level in each availability zone is computed and filtered using KDE. Then, the low-ranked service providers are eliminated, and the correlation between availability zones and services are examined using Formal Concept Analysis Clustering (FCAC) method. Finally, the service selection for the customer is made based on customer ratings and probability of suitability of cloud zone.

Genetic Algorithm (GA) based service selection method proposed by Li Liu optimizes service selection as stated in customer service request [86]. GA uses an incremental approach to obtain the most favorable solutions in wide range spaces. The fitness value is a measure of how good is the selected solution to the problem. GA-based approach applies in three steps: (1) selection, (2) crossover, and (3) mutation to filter services based on fitness value and eliminating poor services. Initially, services are ranked based on fitness value, and then the selection process selects the top-ranked fixed number of services. Next, a pair of services is selected randomly to form service composition and categorized into higher and lower services. Finally, GA chooses a random point to generate better service composition. Table 2.2 exhibits techniques for resource sharing and comparison of agent-based service composition and QoS-based service composition.

2.3   Social Cloud

The advancement in the Internet of Things (IoT) is reconstructing our times into a cyber-physical-social hyperspace and altering what it intends to be social. Prominent cases include smartphones, tablets, and all sorts of wearable gadgets, which are uniting people, both directly and indirectly, through several applications and social networks such as Twitter, Facebook, and WeChat. People tend to establish new connections every day, and it forms a social network. extend the high potential for fostering collaboration among individuals and among groups. This possible collaborative atmosphere is not only suitable for entertainment, but can also contribute significant value to various

TABLE 2.2

Agent-Based Service Composition Versus QoS-Based Service Composition

Selection Criteria	Techniques	Strengths	Challenges
Agent based	Self-coordinating MAS	Easier service discovery	Ensuring quality of services and customer satisfaction
	FSCNP Ontology based recommendation system	Trust worthiness Transaction based price negotiation	
QoS based	KDE based recommendation system	Trust worthiness	Finding similar service
	GA based service selection	More customer satisfaction	Selecting price worthy services

research societies. For this purpose, scientists are frequently employing social networking theories in projects to create groups, share knowledge, advertise their work, and communicate with their companions [11].

Concerning social networking, many techniques are proposed for resource sharing. The Automated Service Provisioning Environment (ASPEN) [9] is one of the methods that provide a combination of Web 2.0, social networks, and the cloud. ASPEN presents applications and shares data inside a business using a social network. It concentrates on the particular research communities which help based on the social features of collaboration. Therefore, the current social networks are possibly suitable platforms for sharing cloud resources in a more general way. There are numerous good examples of cloud and social networks serving synchronically, but in most cases, the cloud is merely entertaining the applications of social networks, for example, a user can develop a Facebook application hosted by Amazon's AWS service.

The SC [11] assists individuals or institutions to share the capacity of their computing resources utilizing virtual machines (VMs) borrowed through the social network. Actors of the SC can share, demand, and utilize VMs from other actors. It is the virtual organization among the group members, but the modern work on the SC is not establishing a proper economic model for, such as incentive mechanisms for benefaction and various economic allocations [3]. The general architecture of the SC is illustrated in Figure 2.3.

There are two different economic models in the SC; firstly, the pricing models and secondly, the bartering models. The pricing models in the cloud resource sharing, deal with the cash for sharing the resources. However, in the case of the bartering models, there is no cash involved in the sharing process.

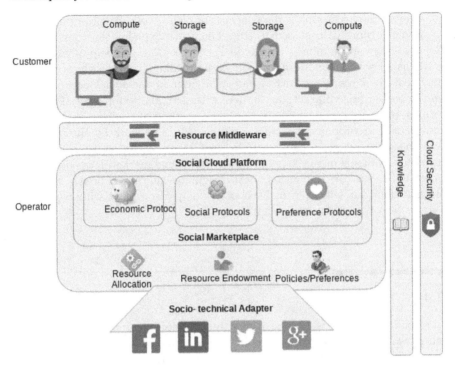

FIGURE 2.3
General social cloud architecture.

There is some work related to the creation of cloud and social network model. A paradigm is designed in [18] for an economic model and an auction-based model in the SC. This model is proposed for sharing storage resources in the cloud.

SC systems utilize financial models, such as auction, bartering, and pricing models. Financial models are classified into two fundamental divisions based on the sharing methods. The cloud industry has effectively used various pricing models for resource sharing [2], but bartering models are not popular because of the trust factor among judicious users. In massively distributed systems, several computational competent resource bartering models are developed in the existence of judicious SC actors. The method in [4] serves a community of colleagues to share cloud resources without including any currency exchange between them. Bartering methods are divided into three distinct classes called as TiT for TaT (BitTorrent) [58], Volunteer Computing [6], and Network of Favors (OurGrid) [8]. BitTorrent aides reducing the capacity on clogged cloud providers by tracing the actions of users. OurGrid encourages the group of people in site sharing. A site provides unused resources based on the centralized preference.

The SC model grouping the concepts of cloud, volunteer computing, and social networking is discussed in [19]. This work explains the resource trading of storage space among friends in Facebook. The problem of freeloading is discussed with the help of a credit-based dealing strategy where the users may rent their resources with a particular user in the social network based on the posted price market, or compete within an auction market. The forum formed around a social group encourages a method of trading, exchanging of resources, and various concurrent auctions [33]. The work by the authors of [18] created a dynamic cloud environment by using the mutual connection naturally available in the social network. The social marketplace concept proposed in the work by the authors of [19] regulates sharing demonstrated social storage cloud implemented in Facebook. Comparison of existing works with various factors is given in Table 2.3.

2.4   Mobile Cloud

In recent times, there is a massive growth in wireless communication networks, where various models of smart mobile devices and lots of smart apps are in use. Mobile devices face the power and resource limitation problems because of energy-hungry and computation-intensive apps. The cloud resources in this context are not only engaged in computation, but also processing, storage, and communications [87]. Despite the rapid growth in the hardware technologies, the problem of resource scarcity persists due to the increasing computation need of the users. Meanwhile, cloud technologies are moving fast in giving an excellent experience to the users in virtualization and sharing the resources. Solving the problem of resource limitation in the mobile devices owns many works by researchers which discuss the well-organized computation offloading mechanism for portable devices to remotely available cloud resources or closely available computing resources called cloudlets. The general computing architecture with cloudlet is depicted in Figure 2.4. There are few of the literature which discusses the various techniques of mobile cloud computing (MCC) [29,41,77].

MCC can play an essential role in providing access to a mobile device for offloading considerable computational responsibilities to the connected cloudlet servers. It ensures the QoS demands of the mobile device users. However, the connection between the mobile access point and the mobile device is slightly irregular with the unstable signal strengths. Additionally, the resource heterogeneity of the cloudlet and the massive data application requirements creates further problems in running the best possible code [24].

Some of the related works that have proposed to support the MCC are given in this section. The works by the authors of [10,29,48,56,59,90] discuss the idea of clustering mobile devices and linking them to the remote cloud.

TABLE 2.3
Related Work Comparison

Methods	Resource	Uses	Integration	Allocation
Social Compute Cloud [14]	Compute	Authentication, social graph extraction	Facebook API, clearinghouse, Django Social Auth plugin	Matching the preference
Social Storage Cloud [19]	Storage	Authentication, social-graph extraction	Integrated Facebook application	Economic - pricing or auction
Social Content Delivery Network (S-CDN) [43]	Storage	Authentication, and social-graph extraction	Facebook API and coauthorship network	Social network analysis
FriendBox [30]	Storage	Authentication, and social-graph extraction	Facebook API	Allocate equally among friends resources
Subdivision Social Cloud [3]	Compute	Authentication, social graph extraction	Integrated Facebook application	Economic - bartering
Social Cloud [57]	None	Social-graph extraction	None	Scheduling-based model
Cycle Sharing in Social Networks (CSSN)[9]	Compute	Social graph extraction, create a social network for message transport	Facebook API	Social postings

(Continued)

TABLE 2.3 (*Continued*)
Related Work Comparison

Methods	Resource	Uses	Integration	Allocation
Multi-community-cloud collaboration (MC3) [9]	Compute	Collaboration across community clouds with social networks	None	Social network analysis
CONFINE [62]	Compute	Ad hoc extension to existing network topologies	None	Scheduling-based techniques and community economies
Social Community Clouds [62]	Compute/data	Authentication and social-graph extraction	Federated cloud scenario using the CometCloud system	Economic (price/reputation)
Cybernetics Social Cloud [16]	Compute/data	Social-graph extraction	Facebook Graph API, hybrid Cloud, and BOINC	Resource sharing between communities.
Facebook-based cloud resource sharing [4]	Compute	Authentication, social graph extraction	Container based CRB model- Facebook API	Economic (bartering)
Altruistic Device Sharing [37]	Data/Power	Authentication, social-graph extraction	Mobile network users and their social closeness	Economic (bartering)

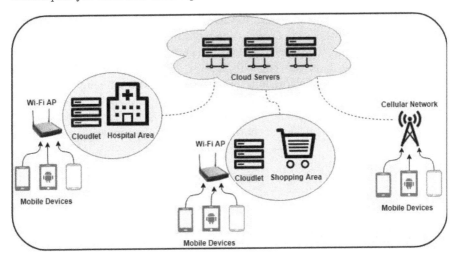

FIGURE 2.4
Architecture of mobile cloud.

Fernando et al. [28] tested the scenario of MCC to utilize local resources to resolve the resource scarcity. A concept of cooperative computing platform named as "Transient cloud" is projected in [56]. This work allows close-by devices to form a network and offer a variety of resources as a service to the actors in the system—tasks assigned to the devices done by the Hungarian technique. Li et al. introduced a best possible solution to distribute the functions to the devices. Zhou et al. [90] designed a model that considers several cloud resources like mobile ad hoc cloud, and cloudlet to provide a MCC. Recently, mobility aware best possible resource provision method, Mobi-Het proposed for remote data-intensive process execution in the mobile cloud with high efficiency and reliability [24]. A broker-based double-sided bidding mechanism for the mobile cloud is introduced in [68]. Few of the use cases of MCC are Smart City [24] and Health Care [35].

2.4.1 Incentive Mechanisms in Mobile Cloud

Although the MCC previously discussed is promising, the proper incentive mechanism encourages the users to share their resources. The following incentive methods are available for MCC: (1) mechanism, (2) bidding, and (3) crowdsourcing. The literature for the incentive mechanisms is given in this section. A fundamental method for auction method is Vickrey-Clarke-Groves (VCG) auction [31,60,73]. There are lots of variation emerged from this primary method, for example, Vickrey-based double auction [60]. McAfee proposed a double auction method [53]. There are few other methods like truthful double auction mechanism (TASC) [80], mClouds [56], CellCloud [59], optimal relay assignment algorithm [63], the maximum weighted matching algorithm [75], and the incentive mechanism based on all-pay auctions [51] proposed in the literature. Double-sided bidding method for mobile cloud resource allocation is proposed in [68]. Incentive mechanisms using crowd-sourcing technique is referred in [20,42,51,81]. Finally, some of the resource allocation optimization algorithms are studied in optimal relay assignment algorithm [63], maximum weighted matching algorithm [75], platform-centric model and user-centric model [81], and optimal incentive mechanism [42]. Thus the mobile cloud resource sharing has lots of open market in the future to the end users.

2.5   Cloud Manufacturing

Cloud intruded into every field. In the manufacturing world, there is a constant pressure to reduce the cost, improve the quality, optimal usage of manufacturing resources, and reduce the impact on the environment. The adoption of has brought many innovations in the manufacturing field. Cloud Manufacturing (CM) enables us to share and use resources optimally with the help of resource sharing. The architecture diagram for CM is shown in Figure 2.5.

CM, irrespective of IT resources and manufacturing resources, and based on its ability, can be classified into seven types which are shown in Figure 2.6 [69].

2.5.1   Resource Sharing in CM

Different manufacturers collaborate to share their capabilities and resources. Two different approaches are used to schedule the jobs when multiple jobs $j_1, j_2, \ldots, j_n$ are placed in the group of machines $m_1, m_2, \ldots, m_n$: (1) perform the jobs in the same order; (2) jobs performed in any order depending on the availability of resources. It is solved by (GA) using various arguments, such as processing time needed, material setup time needed, etc. A service-oriented cloud-based architecture is proposed to solve flow shop scheduling [34].

Complex network-based Cloud Manufacturing Platform (CMP) diffusion model by the authors of [17] was proposed to improve resource sharing. CMP uses theoretical analysis, simulation, and evaluation to match the supply and demand, which facilitate better resource sharing.

In the CM scenario, the resource requester has different choices of providers. The Gale-Shapley algorithm is used to pair the resource requester and provider based on preference degree. A utility function calculates the preference degree based on , price, execution time, reliability, availability, etc. The Gale-Shapley algorithm uses a preference metrics to match the providers [50].

CM tasks are of two types: (1) single resource-based task; and (2) multi-resource based task. Optimal service selection in single resource-based task

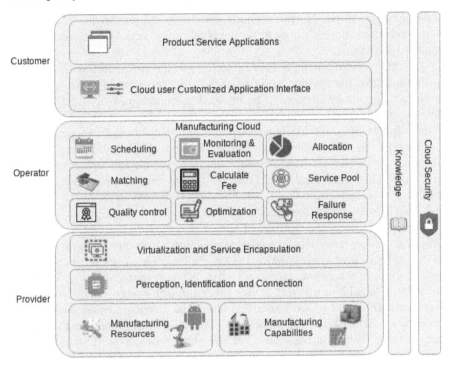

FIGURE 2.5
Cloud manufacturing architecture.

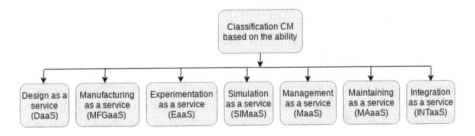

FIGURE 2.6
Ability-based resource sharing in CM.

is simple and straight. In a broad scale, Service Composition and Optimal Selection (SCOS) for multi resource-based task with various constraints and objectives is difficult. Intelligent algorithms are solving multiobjective problems, but they increase the number of iterations and fails to provide efficient resource sharing. FC-PACO-RM, a parallel algorithm [70] is proposed to provide efficient resource sharing decision.

Data-driven smart manufacturing proposed [71] to improve the resource utilization among SMEs. Throughout the product life cycle, a massive amount of data get generated. The proven data analysis techniques are used to envision better resource sharing. Also, it supports proactive maintenance and quality control by learning.

In CM usage of robots is inevitable for specific tasks. Robots need to be positioned into the specific areas to carry out the assigned tasks. Task still is challenging in CM robots. The optimal methods, such as Load Balance Scheduling (LBS), Cost Minimized Scheduling (CMS), and Delay Minimized Scheduling (DMS) provide optimal performance, cost, and processing time. Location of the robot influence in the overall cost. A CM scheduling model is proposed for better resource sharing using the optimal deployment of robots and task scheduling methods. The Location Aware Deployment (LA-DP) reduces the transportation cost, and Function Balanced Deployment (FB-DP) reduces the overall processing time [47].

In CM paradigm the traditional approaches for focus on optimal resource selection. The optimal resource sharing is adopted to achieve operational efficiency. The hybrid energy-aware resource allocation algorithm proposed in [88] to support the demanders uses energy efficient manufacturing. It follows three steps to achieve it. First, it uses fuzzy similarity degree for selecting correct providers based on QoS and then filters the providers based on customer constraints. Second, the non-dominated sorting genetic algorithm (NSGA-II) optimizes the selection based on QoS and energy awareness. Lastly, TOPSIS algorithm is used to select the optimal provider.

Integration of cloud with the manufacturing never changes the dynamic behavior of the cloud. The traditional resource sharing models have a specific problem called semantic heterogeneities. A semantic model is proposed to match the dynamic nature of the cloud. The semantic model enables the flexible access and manipulation of resources [78].

In a manufacturing environment, the production cycle goes like a. The workflow needs a resource to finish a task, and that resource may not be available in a real production environment when it is in a shared environment. The resource sharing and utilization improved if that is managed well. Most existing methods have failed to use the temporal relationship between the resources, which creates an in-efficient utilization. Hence a Resource Service Chain Composition Algorithm (RSCCA) was proposed [46] to build a temporal relationship among the resources to improve resource utilization.

The primary challenge in CM is envisioning the intelligent resource perception and access. IoT plays a significant role in performing intelligent-perception in the shared environment. Identification of various manufacturing resources, services, and their relationship is essential in resource management. IoT-based framework with five layers was proposed for intelligent-view and access to resources of manufacturing [72]. The intelligent-resource perception helps to identify the present state of a resource, which improves the efficiency in sharing.

Mass production reduces the cost of a product in manufacturing. The cloud 3D printing system integrates various 3D printing facilities across the city to reduce manufacturing cost. The cloud 3D receives the order from the demander and distributes it to multiple locations. Location-aware printing helps to finish the work effectively. The existing methods have two problems: (1) to minimize the makespan; and (2) shortest total length path. An agile and ubiquitous additive manufacturing system has been proposed to address the problems mentioned above [21].

2.6 Vehicular Cloud

Vehicle employed with smart devices forms vehicle ad hoc networks (VANET). An enlargement of VANET is VC. Vehicle cloud uses resources like processing, storage, and cyberspace to take a decision which improves traffic management, smart parking, and road safety. Amazingly the high-tech cars are using VCC. The idle resources in the VCC are pooled together and used when their resource demand goes high [54].

2.6.1 Vehicular Cloud Architecture

VCC architecture is visualized as a layered architecture. The VCC architecture is majorly classified into three primary layers, namely inside-vehicle, communication, and cloud [76]. The VCC architecture is shown in Figure 2.7.

2.6.1.1 Inside-Vehicle Layer

The inside-vehicle layer uses smart sensors to observe the health, mood and, STP (Standard Temperature and Pressure) inside the vehicle. It also observes the driver behavior to find his or her unconsciousness and plans. Gathered sensor data is sent to the cloud for further processing.

Roadside units (RSUs) and On-Board Units (OBUs) have transceivers and transponders. The RSU manages OBUs of traveling vehicles via the DSRC technology. DSRC system acquires the needed traffic information, speed, and vehicle location in an instant time. The OBUs transport data through mobile devices.

2.6.1.2 Communication Layer

The communication layer has two components: (1) the vehicular-to-vehicular (V2V) systems via DSRC, and (2) vehicle-to-infrastructure (V2I). V2V recognizes an unusual behavior like exceeding the speed limit, shifting of direction dramatically, or detecting a mechanical failure in the vehicle. On detecting unusual behavior, an alert message is sent to the cloud and nearby vehicles. Alert messages include the geographical location, acceleration, direction of the vehicle, and speed. layer transfers the functional data among vehicles,

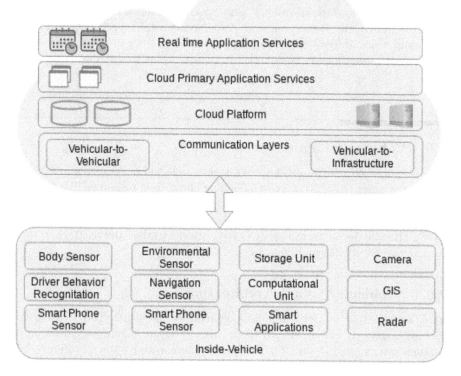

FIGURE 2.7
Vehicular cloud computing architecture.

infrastructures, and the cloud over 4G, internet, or satellite. It ensures vehicle safety on roadways automatically.

2.6.1.3   Cloud Layer

The cloud layer has two components: (1) platform, and (2) application services. Government and related agencies create reports and perform investigation using the data stored in the cloud. The cloud platform has two subcomponents: (1) storage, and (2) processing. Data collected by IVL is moved to the storage and compute component performs a task for effective management of traffic, smart parking, and multimedia services. Application services component incorporate Real-Time Application (RTA) and Cloud Primary Application Services (CPAS). Vehicle drivers access RTA to identify inside and outside human activity with the help of CPAS. CPAS comprises of services, such as NaaS, CaaS, STaaS, INaaS, and ENaaS [45,89].

The components of the CPAS layer are described in Table 2.4.

TABLE 2.4

Primary Application Service Components

Components of CPASL	Functions
NaaS	Shares the internet with other vehicles
STaaS	Storage for vehicles when additional storage needed to run their applications over a network
CaaS	Provides a list of services such as traffic information, road condition, driver safety, and parking availability
Computing	Aggregate the idle computer and storage among parked vehicles and offer it as a service to the demander
Pictures on a wheel	Based on the customer demand takes the photo or video of the landscape and share it
INaaS	Set of information such as road conditions, emergency situation, etc. shared to the drivers for a safe journey
ENaaS	Advertisements and promotions delivered in the screen of the car as like beside the road

2.6.2 Resource Sharing in Traffic Management

Good traffic management leads to smooth transportation. Because of the increase in number of cars on the streets, traffic management became a complicated process. One of the most important things to overcome this problem is to share required resources rather than allocating excess resources, hence utilization of the resources can be improved [52].

Consider the following scenario, a cricket match is scheduled in the city. Many people enter and leave the city to watch the match simultaneously, which causes a traffic jam. Various investigations are carried out to address the above problem based on VANETs and IT system, but failed to provide a traffic moderation plan. Hence, it is necessary to find a solution, which enables the vehicles to pool resources and create a powerful super-computer that, reschedules the traffic lights that regulate the traffic without congestion [23].

2.6.3 Resource Sharing in Multimedia Services

The driving recorder is used in Intelligent Transport Systems (ITS) to record the video of the vehicle. The recorded videos are pushed to the cloud in real time. Pushing multiple videos from multiple Driving Recorders Video (DRVs) creates a competition for the wireless connectivity for uplink. A solution to the above problem comprises of two objectives: (1) assignment of bit rate, and (2) real time slot assignment. The objectives (1) and (2) solved by utility-based priority function was proposed in [39].

2.6.4   Resource Sharing in Smart Parking

(SPS) play a vital role in smart cities. Effective management of traffic and parking increase the satisfaction of commuters. The intelligent car parking systems are getting a lot of recognition and momentum.

The Smart Car Parking System (SCPS) directs vehicles toward the best path for parking in the city using vehicular road sensing and intelligent display. VCC mixes MCC, and vehicular networking (VN) to create better communication and computing. The applications like Netflix is used in vehicular clients. Also, it is used in SCPS. However, the streaming nature of Netflix-like apps brings jitter and delay, which, affects the SPCS. By using technology advancements like fog and edge, the SPCS solves the traffic-related issues [38].

To reduce the parking-related problems, SPCS checks for a vacant slot and then directs the vehicle to the parking if there is a space. Car theft is avoided by using a fingerprint scanner and camera in the parking lot. Introduction of fog improves the of SPCS [7].

2.6.5   Vehicle-Assisted Cloud Computing (VACC) for Smartphones

Even though standard resource exists in the smartphones to process its applications, it needs more processing power based on the apps used such as gaming and mobile data analytics. The sudden resource demand for a smartphone is provided by the Vehicle assisted cloud computing (VACC).

VACC comprises of a central cloud, a cloudlet, and a VC. In VACC architecture, MAN connects the components which need network connectivity. The VC is connected to a cloudlet for resources. Smartphones are connected to the cloudlets through mobile networks or Wi-Fi. The cloud nodes are heterogeneous based on their capacity and resources. The central cloud has large storage and high computational power and less communication rate because it is connected to WAN. The cloudlet has high communication rate, less storage, and less computation [67].

When a smartphone needs an extra resource for its application first, it tries to get from VCC. If VCC does not have the required resources, next, it tries to get from the cloudlet. If this also fails, then it gets the resource form the central cloud [83].

2.6.6   Resource Sharing in Vehicular Fog Computing

The smart vehicles possess decent devices for computation and communication. Vehicular Fog Computing (VFC) is a collaboration of distinct nearby edge devices to perform computation and communication. The nearby moving vehicle can communicate and share the resources with each other to form a

TABLE 2.5

Comparison of Resource Sharing Techniques in VCC

Techniques	Resource Utilization	Response Time	Integration	Security/ Privacy
Traffic Management in VCC	Medium	High	Complicated	Medium
Multimedia Services	Low	More short	Easy	Low
Smart Parking	Medium	Short	Medium	Medium
Vehicle-Assisted	Low	High	Medium	Medium
Vehicular Fog	Very High	High	Complicated	High

good computing power. The parked vehicles can also establish communication for computation [5,36,85]. Table 2.5 shows the comparison of resource sharing techniques in VCC.

2.7 Green Cloud

Green cloud computing enables us to use computer resources in an environmentally responsible way. It is possible by redesigning the data center to minimize power consumption, resource utilization, and the usage of natural resources. Cloud data centers are also adopting green computing to save energy, reduce cost, and save environment. In Computing (GCC), plays a vital role [61].

2.7.1 Resource Sharing in GCC

In cloud data centers thousands of physical machines are used to power up user's VM. consolidation is the process of moving the VMs to the few physical machines, which reduces the consumption of energy. In cloud data center VM consolidation is used. Practically the VM consolidation is an NP-hard problem. Ant Colony based VM consolidation minimizes the energy utilization while maintaining the [27].

The main focus of GCC is using less energy in cloud data center. A clonal selection algorithm [64] proposed to reduce the energy consumption, time, and cost. A scheduling algorithm based on multi-step heuristic proposed in [13] to reduce the CO_2 emission and improve the resource utilization. When the customer request for a resource with a deadline, the request is assigned to one of the data centers, which has less environmental effects without compromising the deadline.

Adaptive energy-aware VM migration technique proposed by Zhang et al. decreases energy utilization and service level agreement (SLA) violation in green cloud computing [79]. The data center performance increases because the technique proposed by Zhang et al. takes four parameters (number of host shutdowns, number of SLA violations, number of VM migration, and energy consumption) while selecting a host for placing new VMs. For the placement of new VMs, it monitors the data center and saves energy. The technique mentioned above uses a central controller (CC), local controller (LC), and VM controller (VMC). LC monitors CPU utilization and VM in the local host and sets any one of the statuses such as under-load, overload, and normal. CC collects statistics from LCs, maintains the overall status of the data center, and has a plan for deployment of the new VMs. VMC migrates and resizes the VMs. The earlier energy-aware VM selection techniques do not consider SLA violation and network bandwidth that reduces the performance of the data center.

Locust-inspired scheduling technique proposed by Kurdi et al. decreases energy consumption by applying Locusts behavior [44]. It places new VMs on the host by executing three phases (mapping, consolidation, and migration). In the mapping phase, an entered VM request is received and assigned to the server with enough resources. Consolidation phase consolidates VMs for reducing the count of servers active. During the migration phase, VMs from underload server is moved to other servers for reducing the number of running servers. LACE algorithm is applied in both mapping and consolidation phases.

2.8   Resource Scheduling Using Meta-Heuristic Techniques

Many works have been done in the area of resource scheduling using meta-heuristic techniques. There are many types of meta-heuristic techniques used in resource scheduling in the cloud computing, such as the Ant Colony Optimization (ACO), Genetic Algorithm (GA), and Particle Swarm Optimization (PCO). ACO is used to optimize the selection of pooled resources. It is mainly used in the VC and SCC. The GA is used to select the quality service providers based on QoS parameters. It is majorly used in task distribution in the CM. The PSO is majorly used in workflow processing and service selection in the cloud. Application of PSO in CM improves the efficiency of the manufacturing process to the next level [40].

Future Scope

In this chapter, the benefits of resource sharing techniques in the cloud computing platform and challenges were discussed. However, some open issues exist that can be a beginning point for future research work. In a environment, selecting a trustworthy and high-quality provider, neglecting selfish providers, and creating transaction histories for new providers are the real challenges. Artificial intelligence techniques can be used to get better performance than using traditional federation management techniques. Numerous difficulties emerge while moving from a cloud model to a SC model. Ensuring trust and long-term service in the SC is a challenging task. Scalability, maintenance, privacy, and security are the other concerns in the SC. Incentive-based resource sharing can solve durability and trustworthiness. Currently, extensive research is going on the areas mentioned above. Open problems are present in the MC resources sharing techniques that can be addressed by solving congestion issues in mobile communication and improving the energy-efficient transmission. The future CM can be constructed by the sustainable manufacturing process to reduce the impact of climate and energy. CM can be improved by incorporating efficient, automated, energy-aware techniques which maximize the production and minimize the waste at a reduced cost. In VCC, many areas are still unexplored for researchers such as security, privacy, responsive resource sharing, highly reliable data access, the mobility of resources sharing, unbalanced message links. Furthers VCC can be extended to intelligent VC computing like driverless vehicles, machine learning based traffic control and route prediction system. Scheduling and resource management play a vital role in GCC. In GCC, environment-aware computing is made possible by energy-aware VM allocation, scheduling, and resource management. Incorporating machine learning techniques in scheduling and resource management can bring better results.

Summary

This chapter states the different resource sharing techniques. It compares various resource sharing techniques from the literature and compares their strengths and challenges. Moreover, the revenue generation models for resource sharing are discussed. The use cases of cloud resource sharing using the mobile, social network, and vehicular network are elaborated. Also, the effect of the green cloud for sustainable development is also discussed.

Bibliography

[1] https://collaborate.nist.gov/twiki-cloud-computing/bin/view/Cloud Computing/CloudPWGFC Vocabulary.

[2] Asmaa Abdallah and Xuemin Shen. Lightweight security and privacy preserving scheme for smart grid customer-side networks. *IEEE Transactions on Smart Grid*, 8(3):1064–1074, 2017.

[3] Zahra Ali, Raihan Ur Rasool, and Peter Bloodsworth. Social networking for sharing cloud resources. *Proceedings of 2nd International Conference on Cloud and Green Computing and 2nd International Conference on Social Computing and Its Applications, CGC/SCA 2012*, Xiangtan, Hunan, China, pages 160–166, 2012.

[4] Zahra Ali, Raihan Ur Rasool, Peter Bloodsworth, and Shamyl Bin Mansoor. Facebook-based cloud resource sharing. *Computers and Electrical Engineering*, 66:162–173, 2018.

[5] Arwa Alrawais, Abdulrahman Alhothaily, Bo Mei, Tianyi Song, and Xiaolu Cheng. An efficient revocation scheme for vehicular Ad-Hoc networks. *Procedia Computer Science*, 129:312–318, 2018.

[6] David P. Anderson, Jeff Cobb, Eric Korpela, Matt Lebofsky, and Dan Werthimer. SETI@home: An experiment in public-resource computing. *Communications of the ACM*, 45(11):56–61, 2002.

[7] Elsie Chidinma Anderson, Kennedy Chinedu Okafor, Onwuchekwa Nkwachukwu, and Damian O. Dike. Real time car parking system: A novel taxonomy for integrated vehicular computing. *Proceedings of the IEEE International Conference on Computing, Networking and Informatics, ICCNI 2017*, 2017-January (VCC):1–9, 2017.

[8] Nazareno Andrade, Walfredo Cirne, Francisco Brasileiro, and Paulo Roisenberg. OurGrid: An approach to easily assemble grids with equitable resource sharing. In *Job Scheduling Strategies for Parallel Processing, 9th International Workshop, JSSPP 2003*, Seattle, WA, June 24, 2003, Seattle, WA, USA pages 61–86, 2003.

[9] Nuno Apolónia, Paulo Ferreira, and Luís Veiga. Enhancing online communities with cycle-sharing for social networks. *Computational Social Networks: Tools, Perspectives and Applications*, 9781447140:161–195, 2012.

[10] Cristian Barca, Claudiu Barca, Cristian Cucu, Mariuca Roxana Gavriloaia, Radu Vizireanu, Octavian Fratu, and Simona Halunga. A virtual cloud computing provider for mobile devices. In *Proceedings*

of the 8th International Conference on Electronics, Computers and Artificial Intelligence, ECAI 2016, Ploieşti, Romania, 2017.

[11] Kris Bubendorfer, Kyle Chard, Koshy John, and Ashfag M. Thaufeeg. EScience in the social cloud. *Future Generation Computer Systems*, 29(8):2143–2156, 2013.

[12] Yunus Emre Cakmaz, Omer Faruk Alaca, Caglar Durmaz, Berkay Akdal, Baris Tezel, Moharram Challenger, and Geylani Kardas. Engineering a BDI agent-based semantic e-barter system. *2nd International Conference on Computer Science and Engineering, UBMK 2017*, Antalya, Turkey, pages 1072–1077, 2017.

[13] Fei Cao and Michelle M. Zhu. Energy efficient workflow job scheduling for green cloud. *Proceedings of IEEE 27th International Parallel and Distributed Processing Symposium Workshops and PhD Forum, IPDPSW 2013*, pages 2218–2221, 2013.

[14] Simon Caton, Christian Haas, Kyle Chard, Kris Bubendorfer, and Omer F. Rana. A social compute cloud: Allocating and sharing infrastructure resources via social networks. *IEEE Transactions on Services Computing*, 7(3):359–372, 2014.

[15] Antonio Celesti, Francesco Tusa, Massimo Villari, Antonio Puliafito, Contrada Dio. How to enhance cloud architectures to enable cross-federation. *2010 IEEE 3rd International Conference on Cloud Computing*, Miami, FL, 2010, pages 337–345.

[16] Victor Chang. A cybernetics social cloud. *Journal of Systems and Software*, 124:195–211, 2017.

[17] Geng Chao, QU Shiyou, Xiao Yingying, Wang Mei, and Shi Guoqiang. Diffusion mechanism simulation of cloud manufacturing complex network based on cooperative game theory. *Journal of Systems Engineering and Electronics*, 29(2):321–335, 2018.

[18] Kyle Chard, Simon Caton, Omer Rana, and Kris Bubendorfer. Social cloud: Cloud computing in social networks. *Proceedings of 2010 IEEE 3rd International Conference on Cloud Computing, CLOUD 2010*, Miami, FL, USA, pages 99–106, 2010.

[19] Kyle Chard, Kris Bubendorfer, Simon Caton, and Omer F. Rana. Social cloud computing: A vision for socially motivated resource sharing. *IEEE Transactions on Services Computing*, 5(4):551–563, 2012.

[20] Cai Chen and Yinglin Wang. SPARC: Strategy-proof double auction for mobile participatory sensing. In *Proceedings of 2013 International Conference on Cloud Computing and Big Data, CLOUDCOM-ASIA 2013*, Fuzhou, China, 2013.

[21] Tin-chih Toly Chen and Yu-cheng Lin. Full length Article A three-dimensional-printing-based agile and ubiquitous additive manufacturing system. *Robotics and Computer Integrated Manufacturing*, 55(July 2018):88–95, 2019.

[22] Sebla Demirkol, Sinem Getir, Moharram Challenger, and Geylani Kardas. Development of an agent based e-barter system. In *2011 International Symposium on Innovations in Intelligent Systems and Applications*, Istanbul, Turkey, pages 193–198. IEEE, June 2011.

[23] Mohamed Eltoweissy, Stephan Olariu, and Mohamed Younis. Towards autonomous vehicular clouds: A position paper (Invited paper). *Lecture Notes of the Institute for Computer Sciences, Social-Informatics and Telecommunications Engineering*, 49 LNICST:1–16, 2010.

[24] Asma Enayet, Md Abdur Razzaque, Mohammad Mehedi Hassan, Atif Alamri, and Giancarlo Fortino. A mobility-aware optimal resource allocation architecture for big data task execution on mobile cloud in smart cities. *IEEE Communications Magazine*, 56(2):110–117, 2018.

[25] Jaeyong Kang and Kwang Mong Sim. Ontology-enhanced agent-based cloud service discovery. *IJCC* 5:144–171, 2016.

[26] Eduardo de Lucena Falcão, Francisco Brasileiro, Andrey Brito, and José Luis Vivas. Enhancing fairness in P2P cloud federations. *Computers & Electrical Engineering*, 56:884–897, 2016.

[27] Fahimeh Farahnakian, Adnan Ashraf, Tapio Pahikkala, Pasi Liljeberg, Juha Plosila, Ivan Porres, and Hannu Tenhunen. Using ant colony system to consolidate VMs for green cloud computing. *IEEE Transactions on Services Computing*, 8(2):187–198, 2015.

[28] Niroshinie Fernando, Seng W. Loke, and Wenny Rahayu. Dynamic mobile cloud computing: Ad hoc and opportunistic job sharing. In *Proceedings of 2011 4th IEEE International Conference on Utility and Cloud Computing, UCC 2011*, Victoria, NSW, Australia, pages 281–286, 2011.

[29] Niroshinie Fernando, Seng W Loke, and Wenny Rahayu. Mobile cloud computing: A survey. *Future Generation Computer Systems*, 29(1):84–106, 2013.

[30] Raul Gracia-Tinedo, Marc Sanchez-Artigas, and Pedro Garcia-Lopez. F2Box: Cloudifying F2F storage systems with high availability correlation. *Proceedings of 2012 IEEE 5th International Conference on Cloud Computing, CLOUD 2012*, Honolulu, HI, USA, pages 123–130, 2012.

[31] Theodore Groves. Incentives in teams. *Econometrica*, 41:617–631, 1973.

[32] J. Octavio Gutierrez-Garcia and Kwang Mong Sim. Agent-based cloud service composition, *Applied Intelligence*, 38(3):436–464. 2013.

[33] Fei Hao, Geyong Min, Jinjun Chen, Fei Wang, Man Lin, Changqing Luo, and Laurence T. Yang. An optimized computational model for multi-community-cloud social collaboration. *IEEE Transactions on Services Computing*, 7(3):346–358, 2014.

[34] Petri Helo, Duy Phuong, and Yuqiuge Hao. Cloud manufacturing – Scheduling as a service for sheet metal manufacturing. *Computers & Operations Research*, 110: 208–219, 2019.

[35] Doan B. Hoang and Lingfeng Chen. Mobile Cloud for Assistive Healthcare (MoCAsH). *Proceedings of 2010 IEEE Asia-Pacific Services Computing Conference, APSCC 2010*, Hangzhou, China, pages 325–332, 2010.

[36] Xueshi Hou, Yong Li, Min Chen, Di Wu, Depeng Jin, and Sheng Chen. Vehicular fog computing: A viewpoint of vehicles as the infrastructures. *IEEE Transactions on Vehicular Technology*, 65(6):3860–3873, 2016.

[37] Yuichi Inagaki and Ryoichi Shinkuma. Shared-resource management using online social-relationship metric for altruistic device sharing. *IEEE Access*, 6:23191–23201, 2018.

[38] Zhanlin Ji, Ivan Ganchev, Máirtín O'Droma, Li Zhao, and Xueji Zhang. A cloud-based car parking middleware for IoT-based smart cities: Design and implementation. *Sensors (Switzerland)*, 14(12):22372–22393, 2014.

[39] Ming Kai Jiau, Shih Chia Huang, Jenq Neng Hwang, and Athanasios V. Vasilakos. Multimedia services in cloud-based vehicular networks. *IEEE Intelligent Transportation Systems Magazine*, 7(3):62–79, 2015.

[40] Mala Kalra and Sarbjeet Singh. A review of metaheuristic scheduling techniques in cloud computing. *Egyptian Informatics Journal*, 16(3):275–295, 2015.

[41] Abdul Nasir Khan, Laiha Mat Kiah, Samee U Khan, and Sajjad A Madani. Towards secure mobile cloud computing: A survey. *Future Generation Computer Systems*, 29(5):1278–1299, 2013.

[42] Iordanis Koutsopoulos. Optimal incentive-driven design of participatory sensing systems. In *Proceedings of IEEE INFOCOM*, Turin, Italy, 2013.

[43] Kai Kugler, Kyle Chard, Simon Caton, Omer Rana, and Daniel S Katz. Constructing a social content delivery network for escience. *2013 IEEE 9th International Conference on e-Science*, pages 350–356, 2013.

[44] Heba A Kurdi, Shaden M Alismail, and Mohammad Mehedi Hassan. LACE: A locust-inspired scheduling algorithm to reduce energy consumption in cloud datacenters. *IEEE Access*, Beijing, China, 6:35435–35448, 2018.

[45] Euisin Lee, Eun Kyu Lee, Mario Gerla, and Soon Y Oh. Vehicular cloud networking: Architecture and design principles. *IEEE Communications Magazine*, 52(2):148–155, 2014.

[46] Haibo Li, Keith C C Chan, Mengxia Liang, and Xiangyu Luo. Composition of Resource-Service Chain for Cloud Manufacturing. *IEEE Transactions on Industrial Informatics* 12(1):211–219, 2016.

[47] Wenxiang Li, Chunsheng Zhu, Student Member, Laurence T Yang, Lei Shu, Edith C Ngai, and Yajie Ma. Subtask Scheduling for Distributed Robots in Cloud Manufacturing. *IEEE Systems Journal* 11(2):941–950, 2017.

[48] Yujin Li, Lei Sun, and Wenye Wang. Exploring device-to-device communication for mobile cloud computing. In *2014 IEEE International Conference on Communications, ICC 2014*, Sydney, NSW, Australia, pages 2239–2244, 2014.

[49] Haikun Liu and Bingsheng He. F2C: Enabling fair and fine-grained resource sharing in multi-tenant IaaS clouds. *IEEE Transactions on Parallel and Distributed Systems*, 27(9):2589–2602, 2016.

[50] Yongkui Liu, Lin Zhang, Fei Tao, and Long Wang. Resource service sharing in cloud manufacturing based on the Gale–Shapley algorithm: Advantages and challenge. *International Journal of Computer Integrated Manufacturing*, 30(4–5):420–432, 2017.

[51] Tie Luo, Hwee Pink Tan, and Lirong Xia. Profit-maximizing incentive for participatory sensing. In *Proceedings of IEEE INFOCOM*, 2014.

[52] Jayshree Dayanand Mallapur and Renuka Takappa Ambiger. Vehicular traffic management using cloud network app. *International Journal of Science, Technology and Society*, 5(6):203–209, 2017.

[53] Randolph Preston McAfee. A dominant strategy double auction. *Journal of Economic Theory*, 56(2):434–450, 1992.

[54] Khaleel Mershad and Hassan Artail. Finding a STAR in a vehicular cloud. *IEEE Intelligent Transportation Systems Magazine*, 5(2):55–68, 2013.

[55] Haithem Mezni. A multi-recommenders system for service provisioning in multi-cloud environment. *Proceedings of International Workshop on Database and Expert Systems Applications, DEXA*, 2017-Augus, pages 142–146, 2017.

[56] Emiliano Miluzzo, Ramón Cáceres, and Yih-Farn Chen. Vision mClouds: Computing on clouds of mobile devices. In *Proceedings of the Third ACM Workshop on Mobile Cloud Computing and Services - MCS '12*, page 9, 2012.

[57] Abedelaziz Mohaisen, Huy Tran, Abhishek Chandra, and Yongdae Kim. Trustworthy distributed computing on social networks. *Proceedings of the 8th ACM SIGSAC Symposium on Information, Computer and Communications Security*, pages 155–160, 2013.

[58] Giovanni Neglia, Giuseppe Reina, Honggang Zhang, Don Towsley, Arun Venkataramani, and John Danaher. Availability in BitTorrent systems. *Proceedings of IEEE INFOCOM*, pages 2216–2224, 2007.

[59] Shahid Al Noor, Ragib Hasan, and Md Munirul Haque. CellCloud: A novel cost effective formation of mobile cloud based on bidding incentives. In *IEEE International Conference on Cloud Computing, CLOUD*, pages 200–207, 2014.

[60] David C. Parkes, Jayant Kalagnanam, and Marta Eso. Achieving budget-balance with vickrey-based payment schemes in exchanges. In *IJCAI International Joint Conference on Artificial Intelligence*, 2001.

[61] Yashwant Singh Patel, Neetesh Mehrotra, and Swapnil Soner. Green cloud computing: A review on Green IT areas for cloud computing environment. *2015 1st International Conference on Futuristic Trends in Computational Analysis and Knowledge Management, ABLAZE 2015*, pages 327–332, 2015.

[62] Ioan Petri, Javier Diaz-Montes, Omer Rana, Magdalena Punceva, Ivan Rodero, and Manish Parashar. Modelling and implementing social community clouds. *IEEE Transactions on Services Computing*, 10(3):410–422, 2017.

[63] Sushant Sharma, Yi Shi, Y Thomas Hou, and Sastry Kompella. An optimal algorithm for relay node assignment in cooperative Ad Hoc networks. *IEEE/ACM Transactions on Networking*, 19(3):879–892, 2011.

[64] Wanneng Shu, Wei Wang, and Yunji Wang. A novel energy-efficient resource allocation algorithm based on immune clonal optimization for green cloud computing. *Eurasip Journal on Wireless Communications and Networking*, 2014:1–9, 2014.

[65] Kwang Mong Sim. Agent-based cloud computing. *IEEE Transactions on Services Computing*, 5(4):564–577, 2012.

[66] Kwang Mong SIM. Agent-based approaches for intelligent InterCloud resource allocation. *IEEE Transactions on Cloud Computing*, PP(c):1, 2018.

[67] Mujdat Soyturk, Khaza Newaz Muhammad, Muhammed Avcil, Burak Kantarci, Jeanna Matthews. From vehicular networks to vehicular clouds in smart cities. *Smart Cities and Homes: Key Enabling Technologies*, 149–171, 2016 (Chapter 8).

[68] Ling Tang, Shibo He, and Qianmu Li. Double-sided bidding mechanism for resource sharing in mobile cloud. *IEEE Transactions on Vehicular Technology*, 66(2):1798–1809, 2017.

[69] Fei Tao, Lin Zhang, V.C. Venkatesh, Y. Luo, Ying Cheng. Cloud manufacturing: A computing and service-oriented manufacturing model. *Proceedings of the Institution of Mechanical Engineers, Part B: Journal of Engineering Manufacture*, 225(10):1969–1976, 2011.

[70] Fei Tao, Yuanjun Laili, Lida Xu, Lin Zhang, and Senior Member. FC-PACO-RM: A parallel method for service composition optimal-selection in cloud manufacturing system. *IEEE Transactions on Industrial Informatics* 9(4):2023–2033, 2013.

[71] Fei Tao, Qinglin Qi, Ang Liu, and Andrew Kusiak. Data-driven smart manufacturing. *Journal of Manufacturing Systems*, 48:157–169, 2018.

[72] Fei Tao, Ying Zuo, Senior Member, Lin Zhang, and Senior Member. IoT-based intelligent perception and access of manufacturing resource toward cloud. *IEEE Transactions on Industrial Informatics* 10(2):1547–1557, 2014.

[73] William Vickrey. Counterspeculation, auctions, and competitive sealed tenders. *The Journal of Finance*, 16(1):8–37, 1961.

[74] Xiaogang Wang, Jian Cao, and Jie Wang. A dynamic cloud service selection strategy using adaptive learning agents. *International Journal of High Performance Computing and Networking*, 9(12):70–81, 2016.

[75] Douglas B West. *Introduction to Graph Theory: Solution Manual (to Second Edition 2001)*. Number 2001. World Scientific Publishing Company, Chapter 5 Page 143, 2005.

[76] Md Whaiduzzaman, Mehdi Sookhak, Abdullah Gani, and Rajkumar Buyya. A survey on vehicular cloud computing. *Journal of Network and Computer Applications*, 40(1):325–344, 2014.

[77] X Fan, J Cao and H Mao. A survey of Mobile Cloud Computing. *ZTE Communications*, 9(1):4–8, 2011.

[78] Cheng Xie, Hongming Cai, Lida Xu, Lihong Jiang, and Fenglin Bu. Resource service toward cloud manufacturing. *IEEE Transactions on Industrial Informatics*, 13(6):3338–3349, 2017.

[79] Rahul Yadav, Weizhe Zhang, Omprakash Kaiwartya, Prabhat Ranjan Singh, Ibrahim A Elgendy, and Yu-Chu Tian. Adaptive energy-aware algorithms for minimizing energy consumption and SLA violation in cloud computing. *IEEE Access*, 6:1–1, 2018.

[80] Dejun Yang, Xi Fang, and Guoliang Xue. Truthful auction for cooperative communications. In *Proceedings of MobiHoc*, page 1, 2011.

[81] Dejun Yang, Guoliang Xue, Xi Fang, and Jian Tang. Crowdsourcing to smartphones: Incentive mechanism design for mobile phone sensing. *MobiCom 2012*, 2012.

[82] Syeda ZarAfshan Goher, Peter Bloodsworth, Raihan Ur Rasool, and Richard McClatchey. Cloud provider capacity augmentation through automated resource bartering. *Future Generation Computer Systems*, 81:203–218, 2018.

[83] Hongli Zhang, Qiang Zhang, and Xiaojiang Du. Toward vehicle-assisted cloud computing for smartphones. *IEEE Transactions on Vehicular Technology*, 64(12):5610–5618, 2015.

[84] Kan Zhang and Nick Antonopoulos. A novel bartering exchange ring based incentive mechanism for peer-to-peer systems. *Future Generation Computer Systems*, 29(1):361–369, 2013.

[85] Ke Zhang, Yuming Mao, Supeng Leng, Yejun He, and Yan Zhang. Mobile-edge computing for vehicular networks. *IEEE Vehicular Technology Magazine*, 12(June):2–10, 2017.

[86] Miao Zhang, Li Liu, and Songtao Liu. Genetic algorithm based QoS-aware service composition in multi-cloud. *Proceedings of 2015 IEEE Conference on Collaboration and Internet Computing, CIC 2015*, pages 113–118, 2016.

[87] Yang Zhang, Dusit Niyato, and Ping Wang. An auction mechanism for resource allocation in mobile cloud computing systems. In *Proceedings of the 8th International Conference on Wireless Algorithms, Systems, and Applications (WASA'13)*. Springer-Verlag, Berlin, Heidelberg, 76–87, 2013.

[88] H A O Zheng, Yixiong Feng, and Jianrong Tan. A Hybrid Energy-Aware Resource Allocation Approach in Cloud Manufacturing Environment. *IEEE Access*, 5:12648–12656, 2017.

[89] Kan Zheng, Lu Hou, Hanlin Meng, Qiang Zheng, Ning Lu, and Lei Lei. Soft-defined heterogeneous vehicular network: Architecture and challenges. *IEEE Network*, 30(4):72–80, 2016.

[90] Bowen Zhou, Amir Vahid Dastjerdi, Rodrigo N Calheiros, Satish Narayana Srirama, and Rajkumar Buyya. A context sensitive offloading scheme for mobile cloud computing service. In *Proceedings of 2015 IEEE 8th International Conference on Cloud Computing, CLOUD 2015*, pages 869–876, 2015.

3

Swarm Intelligent Techniques for Cloud
Service Provider Selection in a Multi-cloud
Environment

Amany M. Mohamed

Cairo University

Hisham M. Abdelsalam

Cairo University

CONTENTS

3.1   Introduction

Provider selection is one of the most imperative decisions when dealing with cloud computing [1,2]. Comparing providers and deciding which provider is the best represents an important issue to the client/customer and providers as well. Client concerns about selecting provider that best fits its needs, while the providers need to benchmark themselves with others to improve their services [3]. This research focuses on cloud provider selection problem from the client/customer perspective.

Due to the growth of public cloud service offerings and its diversity, selecting the right provider that can satisfy the Quality of Service (QoS) requirements of the customer become an important challenge [4]. The diversity comes from that similar services are offered at different prices and performance levels. Consequently, three main questions should be answered when comparing cloud providers: (1) what are the criteria that will be used to compare cloud providers?, (2) how to measure these criteria?, and (3) how to rank cloud provider based on these criteria? [5].

Cost is considered as an essential factor for comparing cloud providers. Numerous research papers proposed several decision models that were mainly based on cost [3–10]. However, because of the diversity of cloud services offerings, in addition to the increasing number of cloud providers, customer faces a challenge of selecting cloud provider considering various criteria. Consequently, research papers tried to identify the main criteria from customer's perspective [10,11]. In this context, the Cloud Service Measurement Index Consortium (CSMIC) has designed the Service Measurement Index (SMI) to provide a standardization method of measuring and comparing the business service [12].

SMI involves various categories that helps customers to compare different cloud service providers which lead them to consider multiple criteria through the provider selection process. Therefore, to address this problem, previously published research papers proposed multi-criteria decision models. The proposed models were solved using weighted methods, such as Analytic Hierarchy Process (AHP), Ranked Voting Method, or Technique of Order Preference by Similarity to Ideal Solution (TOPSIS). However, the main problem with these methods is the dependency on user experience and preferences which make the selection decision inaccurate. Moreover, weighted methods were used for the purpose of rank providers which means selecting a single provider to satisfy customer's requirements (vendor lock-in problem). Consequently, the customer is forced to accept the provider's capabilities and pricing.

Multi-cloud environment has attracted the attention of the recent research papers—a strategy that assists customers to avoid vendor lock-in problem. Therefore, this chapter presents a general model for selecting providers in multi-cloud environment considering any number of Infrastructure as a service (IaaS) based on two evaluation criteria: cost and performance. The problem was formulated as integer programming and was approved to be a NP-hard problem. Consequently, to solve it, three meta-heuristic algorithms were used: Genetic Algorithm (GA), Harmony Search (HS), and Particle Swarm Optimization (PSO). In order to test and compare the performance of the proposed algorithms, a case study was generated. Results showed improved performance of the PSO compared to the GA and the HS.

The rest of this chapter is organized as follows: Section 3.2 provides related work, while Section 3.3 deals with problem description; it starts with context and proceeds with model formulation. The details of the solution algorithms are presented in Section 3.4. Section 3.5 discusses case study, while Section 3.6 tests the performance of the proposed solution algorithms. Finally, conclusion and future work are given in Section 3.7.

3.2 Related Work

Early papers cited, relevant to this research direction, focused on supporting customer/decision maker (DM) in evaluating only one cloud provider. KhajehHosseini et al., in [6] provided cost modeling that helped DM in obtaining precise cost estimates of running a server infrastructure on the cloud. Then, the idea was extended to include a group of providers; KhajehHosseini et al., in [7] extended their work discussed in [6] in order to use it to compare the costs of different cloud providers and different types of instances. Besides, Li et al., in [3] developed CloudCmp tool to compare cloud providers based on performance and cost and provide this information to the customer to select the right provider. However, all previous researches were restricted to a few cloud providers that dominate the cloud market and the common services offered by them.

As a consequence of the diversity of cloud services offerings and the increasing number of cloud providers, research papers proposed decision models in order to compare different providers considering multiple criteria. Various criteria are considered in the literature, but many of the research papers revolve around cost criteria [3–10]. Furthermore, based on SMI, Garg et al., [5] provided a framework to measure various SMI attributes and rank the cloud services based on it.

As a result of the continually increasing number of cloud service provider's evaluation criteria, research papers depended on user experience and preferences to select the best cloud service provider. Analytic Hierarchy Process (AHP) was the most used method to rank cloud providers [4,5,8–10,14,15]. AHP is a flexible and powerful tool for dealing with complex decision-making. It depends on three principles: (1) decomposition of the problem, (2) comparative judgment of the elements, and (3) synthesis of the priorities. However, when a large number of criteria is considered, applying AHP becomes time-consuming. Another weighted methods were used to solve the provider selection problem. Baranwal and Vidyarthi [16] and Shirur and Swamy [17] proposed Ranked Voting Method, while Upadhyay [18] proposed Technique of Order Preference by Similarity to Ideal Solution (TOPSIS).

All the previously published research work share three common shortcomings: (1) focusing on single-cloud provider perspective; the proposed models and frameworks tried to evaluate different criteria and rank different providers to select only one of them to satisfy customer's requirements and this force the customer to accept the provider's capabilities and pricing, (2) ignoring the constraints for each criterion, and (3) the solution of the problem was based on weighted methods. Recent researches tried to take advantage of multi-cloud to secure customer data getting customer's data storage requirements from multiple cloud providers [19–22].

3.3   Problem Description

3.3.1   Context

Consider a set of autonomous cloud providers. Each provider offers various infrastructure services with different configurations. Each configuration is defined by different parameters. Besides, a client has a set of different infrastructure services to be rented. Based on the type of service, the parameters that defines each service can be numerical or non-numerical. The client needs to satisfy its requirements with minimum cost and maximum performance. The aim of this study is to select the optimal group of cloud providers for all clients' requirements in order to minimize cost and maximize performance subject to a set of constraints.

3.3.2   Model Formulation

For the convenience of formulation, the parameters and variables for provider selection are defined as follows:

$S^{(R)} = S_i^{(R)}|i = 1...N$ a set of N service requirements of client

$S^{(RP)} = S_{ia}^{(RP)}|a = 1,\ldots,A_i$ a set of parameters that define required service i

$P = P_j|j = 1,\ldots,M$ a set of independent cloud service providers

$S_j^{(P)} = S_{kj}^{(P)}|k = 1,\ldots,K$ a set of infrastructure service that offered by provider j

$S_{kj}^{(PP)} = S_{bkj}^{(PP)}|b = 1...B_{kj}$ a set of parameters that define service k that offered by provider j

C_{ij}	cost of required service i when it is rented from provider j
BU	budget of client
PF_{ij}	performance when rented requirement i from provider j
$ART_{ij}^{(P)}$	the provider j response time for the service i
T_{uij}	time between when user u requested for service i from provider j and when it is actually available
n_{ij}	total number of service requests of service i from provider j
$ART_i^{(R)}$	client's maximum acceptable response time
$Suit_{ij}$	value of suitability for service i when rented from provider j
$F_{ij}^{(P)}$	number of essential and non-essential features provided by the service i at provider j
$F_{ij}^{(R)}$	number of essential and non-essential features required by the client for service i

3.3.3   Decision Variable

Given a set of requirements and the set of cloud providers, the decision to be make is which provider to select in order to satisfy each of clients' requirements. So, the decision variable is formulated as integer as follows:

$$X_{ij} = \begin{cases} 1, & \text{if required service i is rented from cloud provider j.} \\ 0, & \text{otherwise.} \end{cases} \quad (3.1)$$

3.3.4   Objective Functions

1. Cost
   Cost is one of the important attributes for IT and the business. It is the most quantifiable metric through which the organization can answer the question whether switching to cloud computing is cost-effective or not [5]. Nowadays, various pricing models/schemes are used by cloud providers, however, "pay-as-you-go" model is the foremost common model used in cloud computing [23,24]. In this study, the cost formulation is based on "pay-as-you-go" model. Majority of services have a constant coefficient of calculation because of the fixed relationship between unit-price and consumption [4]. However, the cost/price of some services will be changed with the increase of

Storage Capacity	GRS
First 1 TB / Month	$0.048 per GB
Next 49 TB (1 to 50 TB) / Month	$0.0472 per GB
Next 450 TB (50 to 500 TB) / Month	$0.0464 per GB
Next 500 TB (500 to 1000 TB) / Month	$0.0456 per GB
Next 4000 TB (1000 to 5000 TB) / Month	$0.0448 per GB
Over 5000 TB / Month	Contact us

FIGURE 3.1
Azure storage service pricing for GRS option.

consumption. In this context, a dynamic function will be used to calculate the service cost. For example, the windows Azure storage service pricing policy for the Geographically Redundant Storage (GRS) option for the Central-US region with the defined cost of $0.01 per 100 K transactions is shown in Figure 3.1.

From the pricing details, we can notice that the unit-price of storage capacity is constant, but the cost will decrease with the increase of amount. Subsequently, it is necessary to set calculation, respectively for each amount range. Thus, given that $S_1^{(PP)}$ is the storage capacity in TB and $S_2^{(PP)}$ is the number of storage transactions, the cost can be formulated as follows:

$$f\left(S_1^{(PP)}, S_2^{(PP)}\right) = \begin{cases} \left(0.048 + \frac{S_1^{(PP)}}{1,000}\right) + \frac{\left(0.01 \times S_2^{(PP)}\right)}{100,000}, \\ \qquad\qquad S_1^{(PP)} \in [0,1] \\ \left(48 + \left(0.0472 \times \frac{S_1^{(PP)}-1}{1,000}\right)\right) + \frac{\left(0.01 \times S_2^{(PP)}\right)}{100,000}, \\ \qquad\qquad S_1^{(PP)} \in (1,50] \\ \left(2360.8 + \left(0.0464 \times \frac{S_1^{(PP)}-50}{1,000}\right)\right) + \frac{\left(0.01 \times S_2^{(PP)}\right)}{100,000}, \\ \qquad\qquad S_1^{(PP)} \in (50,500] \\ \dots \ \dots \end{cases}$$

(3.2)

Consequently, based on the parameters of each service, the cost function can be formulated as follows:

$$C_{ij} = f\left(S_{1j}^{(PP)}, S_{2j}^{(PP)}, ..., S_{B_{kj}}^{(PP)}\right) \qquad (3.3)$$

2. Performance
It is necessary for the clients to understand how well their applications will perform on the different clouds and whether these deployments meet their expectations [5]. Performance is one of the defined categories of SMI. It contains five attributes that are used to measure the features and functions of the provided services. In this

study, two convenient attributes for IaaS services were chosen: Average Response Time (ART) and Suitability (Suit).

(a) Average Response Time

ART is a high-level measure that can be used to quantify service efficiency by calculating the time interval between the service request and the response of the service; how quick the service can be made accessible for utilization [12]. ART, for an existing service, can be calculated using the arithmetic mean (average) of the individual response times over a period of 6 months [12]. ART for required service i when rented from provider j can be determined mathematically using the following equation:

$$\text{ART}_{ij}^{(P)} = \sum_{u}^{U_{ij}} \frac{T_{uij}}{nij} \tag{3.4}$$

From the customer's perspective, an acceptable maximum value for the response time need to be determined, thus the selected provider must meet this value. This constraint is formulated as follows: $\text{ART}_{ij}^{(P)} \geq \text{ART}_{i}^{(R)}$.

(b) Suitability

The client may have fundamental and unnecessary features. Suitability measure is used to answer a question, such as how nearly the provided service matches the clients' needs? CSMIC determine a range from 0 to 10 to express the measure. Consequently, suitability of the required service i that will be rented from the provider j can be presented mathematically as follows:

$$\text{Suit}_{ij} = \begin{cases} 10, & \text{if all the essential features are satisfied} \\ \frac{F_{ij}^{(P)}}{F_{ij}^{(R)}}, & \text{if all essential features are satisfied and some} \\ & \text{of the non-essential features are not satisfied} \\ 0, & \text{if any of the essential features is not satisfied.} \end{cases} \tag{3.5}$$

This measure is customer focused [5] and needs to be performed at the initial selection of the cloud provider [12]. Zero is unacceptable value of suitability, so the suitability of provider must be greater than zero ($\text{Suit}_{ij} > 0$).

In the provider selection process, the provider with the minimum ART will be preferable and the provider with maximum suitability will be preferable. In order to sum the two measures, ART will be multiplied by (-1). Given the weight of each attribute/measure, performance of the service i when rented from the provider j will be calculated according to the following equation:

$$PF_{ij} = w_1^{(r)} \times \text{Suit}_{(ij)} - w_2^{(r)} \times \text{ART}_{ij}^{(P)} \tag{3.6}$$

where $w_1^{(r)}$ is the weight of suitability and $w_2^{(r)}$ is the weight of average response time attribute. Finally, the formulation of provider's selection in multi-cloud environment is presented as follows:

$$\text{Minimize} \sum_{i}^{N} \sum_{j}^{M} C_{ij} \times Xij \qquad (3.7)$$

$$\text{Maximize} \sum_{i}^{N} \sum_{j}^{M} PF_{ij} \times Xij \qquad (3.8)$$

Subject to:

$$\sum_{i}^{N} \sum_{j}^{M} Xij = 1 \qquad (3.9)$$

$$\sum_{i}^{N} \sum_{j}^{M} C_{ij} \times Xij \leq BU \qquad (3.10)$$

$$\text{ART}_{ij}^{(P)} \geq \text{ART}_{ij}^{(R)}, \forall i \in N \qquad (3.11)$$

$$\text{Suit}_{ij} > 0, \forall i \in N \qquad (3.12)$$

$$X_{ij} \in 0,1 \qquad (3.13)$$

The constraint in Equation (3.9) ensures that each required service is rented from only one cloud provider. In Equation (3.10), the total cost cannot be larger than the client budget. In Equation (3.11), the provider response time must meet the client's maximum acceptable response time. The constraint in Equation (3.12) ensures that the suitability of provider must be greater than zero. The last constraint ensures that X_{ij} is a binary variable.

3.4 Solution Algorithms

The multi-cloud provider selection is similar to the multi-item vendor selection (MIVS) problem discussed in [25]; MIVS problem is a problem of procuring a number of different discrete items (N) from different vendors (M) under the assumptions that each item is bought from a single vendor and all vendors offers all items at different costs. MIVS problem is formulated as an integer programming model which is similar to the facility location problem, a well-known NP-complete problem. Given the assumption that each item is bought from a single provider, and the number of candidate solutions is M^N. Consequently, solution space may grow into very large number with the increase in the number of vendors and the number of items to be purchased.

The presented problem is similar to MIVS problem; each of the IaaS requirements is considered as a discrete item and is rented from a single cloud provider. It is formulated as integer programming which is similar to the facility location problem. So, multi-cloud provider selection is NP-complete problem and to solve it three swarm algorithms were applied: Genetic Algorithm (GA), Particle Swarm Optimization Algorithm (PSO), and Harmony Search Algorithm (HS). Due to the difference in the nature of the presented solution algorithms, each algorithm has its own solution representation. All algorithms are common in solution evaluation.

3.4.1 Solution Evaluation

The presented solution algorithms have different solution representation. However, the three algorithms are common in solution representation as shown in Figure 3.2. Thus, for the implementation purpose, a general evaluation function was coded based on the common solution representation.

In the presented model, we have two objective functions: cost and performance.

1. Cost

   Each of the required services has its own function to calculate its cost. For example, the cost function of renting Virtual Machine (VM) service is different from the cost function of renting storage service. Even with the same type of infrastructure service, different cost functions may be used based on the parameters that define each service. For example, for storage service, file storage service and blob storage service have different cost function with different parameters.

   Consequently, in order to calculate the cost of each required service, for a single solution, the first step is determining the corresponding cost function based on the service type. The second step is finding the corresponding cost data for the provider assigned to the required service. Finally, the cost of the required service when renting from the corresponding provider can be measured by

$s_1^{(R)}$	$s_2^{(R)}$	$s_3^{(R)}$	...	$s_N^{(R)}$
3	10	20	...	6

FIGURE 3.2
Common solution representation.

substituting in the service cost function. The previous steps are repeated until all required services are finished. The total cost of the current solution is the summation of the cost of each required service.

2. Performance

   In the presented model, performance has two attributes: suitability and average response time.

   (a) For suitability, it is calculated using Equation (3.5). For simplicity and without loss generality, all the essential features of the required service are assumed to be satisfied by all providers. Thus the suitability of each required service in the current solution is equal to 10 and the total suitability of the current solution is $(10 \times N)$, where N is the number of required services.

   (b) For average response time (ART), each provider claim *ART* for each offered services based on the service type. ART of each required service is the ART of the assigned provider and ART of the current solution is the summation of *ART* of each required service.

   The performance of the current solution is the summation of the calculated value of each attribute multiplied by weighting coefficient.

Note that each attribute has different unit; suitability is represented by numbers ranged from 0 to 10, while ART is measured in seconds. Thus, linear normalization (II) [26] is used to unify the units of attributes. Using Equation 3.14, each attribute is normalized based on its lower bound and upper bound, where r_k is the normalized value of attribute k, $k = 1, 2, \ldots K$, V_k is the calculated value of attribute k, and $v^\sim$ and v^* are lower bound and upper bound of attribute k, respectively

$$r_k = \frac{v_k - v^\sim}{v^* - v^\sim} \tag{3.14}$$

The weighted sum method was selected to convert the presented multi-objective optimization problem to a single-objective optimization by summing the cost and performance function multiplied by weighting coefficients. Note that, the performance of the current solution is normalized, so before applying the weighted sum method, cost is needed to be normalized using linear normalization (II).

3.4.2   Genetic Algorithm (GA)

GA is a heuristic search based on the Darwin's theory of natural evolution. "Survival of the fittest" is the basic concept of GA. As in natural, the strong remains fit, while the weak is removed completely. In the same context, solution with the highest fitness value will be survived during different generations, while the solution with the lowest fitness value will be eliminated [27].

GA is a population-based algorithm. It starts with generating an initial feasible population (set of chromosomes/solutions) randomly or by the use of some heuristics. Fitness function (objective function) is used to measure the fitness of each solution. Then, a new population is created using the three operators of GA: (1) selection/reproduction, (2) crossover, and (3) mutation. The selection operator chooses the fitter solutions; the crossover operator selects two solutions randomly in order to exchange genes and mate them; and the mutation changes some of chromosomes' genes randomly [28]. Figure 3.3 shows pseudo-code of GA.

1. Chromosome Encoding
    For the presented problem, integer encoding representation is suitable. Every chromosome is a string of discrete integer values. Each gene contains information about the provider ID that is potentially assigned to the corresponding client's requirement. Each gene takes values from 1 to M (M, the total number of cloud providers). Chromosome encoding is represented in Figure 3.4

2. New Population
    In order to generate a new population, GA uses three common operators: selection, crossover, and mutation. Selection step involves the

GENERATE random set of chromosomes	// initial population
EVALUATE the fitness of each chromosome	
REPEAT	
SELECT two parents	
CROSSOVER the parents	// with crossover probability
MUTATE the new offspring	
PLACE new offspring in new population	
EVALUATE the new population	
FIND the best	
END IF	
UNTIL stopping criteria	
RETURN the best	

FIGURE 3.3
Pseudo-code of GA.

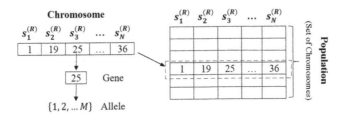

FIGURE 3.4
GA representation.

decision of how to choose two or more parents from the population for crossing. It is an important decision that affects the convergence of GA. According to the Darwin's theory of evolution, the individuals with higher fitness value have more chance to be selected. There are many methods for selection. Some of them depend on the relative fitness value of the individuals, while the others are based on the individuals rank with the population. In this study, roulette wheel selection method was selected. In roulette wheel, the sizes of segments are different based on the relative fitness of each individual. Consequently, the individual with highest fitness will have more probability for selection.

After selecting parent individuals/chromosomes, crossover is used to create new offspring. In this study, single point crossover was applied by selecting crossover point randomly and then the sections of two parents are exchanged. Assuming eight requirements and crossover point equal to three, the parents before and after applying crossover are shown in Figure 3.5.

Finally, in order to maintain diversity within the population, mutation was applied by selecting one gene randomly and generate a new provider ID randomly. As shown in Figure 3.6, assume that the selected gene for mutation is the fifth gene that contains provider with ID = 37. After mutation the fifth gene will contain provider ID = 55 that generated randomly.

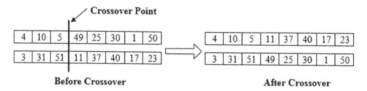

FIGURE 3.5
Single point crossover.

| 4 | 10 | 5 | 11 | 37 | 40 | 17 | 23 | **Before Mutation** |

| 4 | 10 | 5 | 11 | 55 | 40 | 17 | 23 | **After Mutation** |

FIGURE 3.6
Example of mutation.

3.4.3 Particle Swarm Optimization (PSO)

PSO is a population-based stochastic optimization algorithm introduced by Kennedy and Eberhart in 1995 [29]. It inspired by observing the behavior of bird flocking and fish schooling and is mainly based on the guidance of particle to search for global optimal solutions. Each particle represents a potential solution to the problem and has coordinates and rate of change (velocity) in a D-dimensional space. Particles' velocities is one of the important parameters for adjusting the directions of the particle toward their target. Particles' velocities are determined by two factors: (1) their own best experience (cognition part), and (2) the best experience of all other particles (social part).

The original version of PSO is limited in real number space. Therefore, for discrete optimization problem, Kennedy and Eberhart [30] developed a discrete version of PSO. Discrete PSO allows the particle to be composed of binary variable. Besides, to update the particle velocity is transformed into the change of probability, which is the chance of binary variable taking a value one. Figure 3.7 shows the pseudo-code of PSO.

1. Particle Definition

   The solution space of the presented problem is discrete. Therefore, to be able to use a discrete PSO, the solution is represented in different way. As in GA, the initialization of the initial population/particles is generated randomly; for each required service, a random cloud provider is selected. However, to apply the discrete PSO, the generated population is expressed in binary notation. As shown in Figure 3.8, a particle is represented by a matrix with M rows and N columns, where M is the number of cloud providers and N is the number of required services.

   For example, in order to represent that cloud provider with ID = 4 is the candidate provider that will satisfy the first required service $(s_1^{(R)})$, the element at the fourth row and first column will contain 1 and the remaining elements in the first column will

---

INTIALIZE population, velocity, best position,
          global position, and PSO parameters
EVALUATE population
REPEAT
      UPDATE velocity
      UPDATE position
      EVALUATE population
      FIND best position
      FIND global best
UNTIL stopping criteria

FIGURE 3.7
Pseudo-code of PSO.

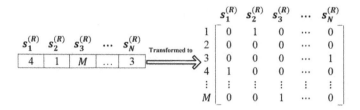

FIGURE 3.8
PSO solution representation.

contain 0 to satisfy constraint (3.9); each required service is rented from a single provider.

2. Generate New Population
   A new population is created by updating the velocity and position of all particles. Velocity is updated as in continuous PSO according to cognition part and social part. Mathematically, using Equation (3.15)

$$v_{id}^{t+1} = wv_{id}^t + \varphi_1 r_1(\text{pbest}_{id}^t - x_{id}^t) + \varphi_2 r_2(\text{gbest}_{gd}^t - x_{id}^t) \quad (3.15)$$

where φ_1 is the cognition learning factor, φ_2 is the social learning factor, r_1 and r_2 are random numbers uniformly distributed in $[0, 1]$, pbest_{id}^t is the individual's i best position found so far at site d, gbest_{gd}^t is the neighborhood best state found so far at site d, x_{id} is the current position at site d of particle i, v_{id}^t is the current velocity at site d of particle i, and w is inertia weight.

The new velocity of each particle is used to update its position by using Equation (3.16) which present the sigmoid function to calculate the probability of x_{id}^t bit taking 1

$$s(v_{id}^t) = \frac{1}{1 + exp(-v_{id}^t)} \quad (3.16)$$

After updating the position of all particles, we need to be sure that new particles/solutions are feasible solutions. As mentioned before, each required service is rented from one provider. Consequently, a simple heuristic is used to check the feasibility of the new particles as follows: for each required service, if more than one provider was assigned, one of them is selected randomly and assign "1" in the corresponding bit and "0" is assigned to the remaining providers. The heuristic for one particle is explained using Figure 3.9. For simplicity, let us assume that client's requirements are three services and ten providers are available to satisfy the requirements. As shown in Figure 3.9, according to the unfeasible particle, the required service $s_1^{(R)}$ can be rented from providers with IDs (3, 4, 6, 7, and 9).

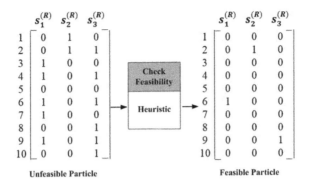

FIGURE 3.9
Feasibility check for one particle.

Thus, a random provider ID is selected from the assigned IDs. For example, let the selected random provider be provider with ID = 6. Consequently, "1" is assigned to the selected provider and "0" is assigned to the remaining providers.

3.4.4 Harmony Search (HS)

HS is a music-inspired algorithm developed by Zong Woo Geem [31] through simulating the musicians' behavior in producing a piece of music with perfect harmony. Skilled musician rely on improvisation through three possible choices: (1) use his memory to play a known piece of music exactly as it is (harmony memory); (2) play a known piece with adjusting the pitch a little (pitch adjusting); or (3) create a piece of music randomly (randomization). Lee and Geem [32] used the same concepts of the three choices of improvisation process to formalize it into quantitative optimization process.

Harmony memory plays an important role to ensure that the best harmonies will be moved to the new harmony memory. The selection is controlled by the harmony consideration rate (HMCR). With the low rate of the HMCR, few best harmonies are selected, whereas with the high rate, almost all harmonies are chosen. Low HMCR results in slow convergence, whereas high HMCR results in poor exploration. Therefore, HMCR are always >70%. For each selected harmonies, pitch adjustment is applied according to pitch adjusting rate (PAR). In order to increase the diversity of the solutions, randomization is applied with probability (100 HMRC) %. Figure 3.10 shows the pseudo-code of the HS algorithm.

1. Harmony Representation
   Harmony representation is similar to chromosome representation discussed in Section 3.4.2 Figure 3.11 shows solution representation in terms of the HS algorithm.

```
DEFINE Harmony Memory Size (HMS), HMCR, PAR
INITIALIZE harmony memory
EVALUATE harmonies
REPEAT
            APPLY harmony memory
            APPLY pitch adjusting
            Apply randomization
            UPDATE harmony memory
        UNTIL stopping criteria
```

FIGURE 3.10
Pseudo-code of the HS algorithm.

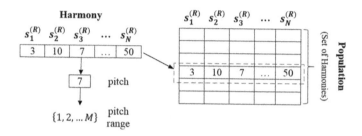

FIGURE 3.11
HS representation.

2. Create New Harmony Memory

   A new harmony memory is created by applying three different options: (1) harmony memory, (2) pitch adjusting, or (3) randomization. Let us assume that we have three required services and ten providers. Given harmony memory size equal to 5, the initial harmony memory is created as shown in Figure 3.12.

   Let HMCR = 75% and PAR = 20%. For the first harmony, random number is generated (HMCRRnd=0.326). HMCRRnd is less than HMCR. Thus, the first value is kept, then another random number is generated to test the probability of changing current pitch value (PARRnd=0.633). PARRnd is greater than PAR, thus

	$s_1^{(R)}$	$s_2^{(R)}$	$s_3^{(R)}$
1	3	1	3
2	4	4	10
3	10	5	3
4	2	2	7
5	5	5	8

FIGURE 3.12
Initial harmony memory.

	$s_1^{(R)}$	$s_2^{(R)}$	$s_3^{(R)}$
1	3	6	7
2	3	4	10
3	10	10	3
4	1	2	7
5	5	5	8

FIGURE 3.13
New harmony memory.

the same value is copied into new harmony memory. For the second pitch, HMCRRnd=0.207 and HMCRRnd < 0.75, then the same value is moved to the new memory. When PARRnd=0.186 and PARRnd < 0.2, then a little change is applied to the pitch value. In this research, the change is suggested to be as follows: generate a new provider ID randomly and add this value to the current pitch value. Assume that the generated provider ID is "5," the new pitch value will be $(1+5 = 6)$. Note that before adding the new pitch value to the new memory, we need to ensure that it is a valid provider ID with the range $[1,M]$, where M is the number of providers. In this example "6" is a valid ID.

For the last requirement, HMCRRnd = 0.78 and HMCRRnd > 0.75, so randomization option is applied. Let the random value be "7." After applying the three options to the first harmony, the new harmony/solution is 3, 6, and 7. By applying the same procedure to the rest of harmonies, the new harmony memory is generated as shown in Figure 3.13.

3.5   Numerical Experiments

For the purpose of testing the performance of the three algorithms, an experimental case study was generated; the most popular cloud providers were scanned and the needed information was collected. Three IaaS services were selected as common service offered by the scanned providers: (1) virtual machine (VM), (2) blob storage (SB), and (3) file storage (SF). Each service is defined by different set of parameters—VM is defined by the following parameters: instance configuration (Conf), operating system (OS), OS type, region (R), while SB is defined by region, capacity, access tier, and redundancy; and SF is defined by region, capacity and redundancy. Ten regions were selected as listed in Table A.1. For VM service 49 common configurations were chosen and six different OS types were selected as presented in Tables 3.1 and 3.2, respectively.

TABLE 3.1

VM Configurations

ID	Cores	RAM	Disk Sizes/GB	ID	Cores	RAM Sizes/GB	Disk
Conf1	1	0.75	20	Conf26	8	16	128
Conf2	1	1.75	40	Conf27	16	32	256
Conf3	2	3.5	60	Conf28	2	14	100
Conf4	4	7	120	Conf29	4	28	200
Conf5	8	14	240	Conf30	8	56	400
Conf6	1	2	10	Conf31	16	112	800
Conf7	2	4	20	Conf32	20	140	1,000
Conf8	4	8	40	Conf33	32	256	512
Conf9	8	16	80	Conf34	64	432	864
Conf10	2	16	20	Conf35	2	28	384
Conf11	4	32	40	Conf36	4	56	768
Conf12	8	64	80	Conf37	8	112	1,536
Conf13	2	7	100	Conf38	16	224	3,072
Conf14	4	14	200	Conf39	32	448	6,144
Conf15	8	28	400	Conf40	64	1,750	2,000
Conf16	16	56	800	Conf41	128	2,000	4,000
Conf17	2	8	16	Conf42	4	32	678
Conf18	4	16	32	Conf43	8	64	13,88
Conf19	8	32	64	Conf44	16	128	2,807
Conf20	16	64	128	Conf45	32	256	5,630
Conf21	32	128	256	Conf46	8	56	1,000
Conf22	64	256	512	Conf47	16	112	2,000
Conf23	1	2	16	Conf48	8	112	1,000
Conf24	2	4	32	Conf49	16	224	2,000
Conf25	4	8	64				

TABLE 3.2

OS Types

ID	OS Name
OS1	Red Hat Enterprise Linux
OS2	SUSE Linux Enterprise
OS3	Windows OS
OS4	Windows with SQL Standard
OS5	Windows with SQL Web
OS6	Windows with SQL Enterprise

In order to calculate the cost and performance of each service, prices and average response times offered by providers were scanned and the absolute minimum and the absolute maximum were collected. Example of the prices intervals per hour for the VM in dollars, storage prices per GB in dollars, and ART for VM and storage services are presented in Tables 3.3–3.5,

TABLE 3.3

Example of the Price Ranges Per Hour for the VM in Dollars

		R1		R2		R3	
		Min	Max	Min	Max	Min	Max
Conf1	OS1	0.078	0.312	0.082	0.328	0.084	0.336
	OS2	0.118	0.472	0.122	0.488	0.124	0.496
	OS3	0.018	0.072	0.022	0.088	0.024	0.096
	OS4	0.418	1.672	0.422	1.688	0.424	1.696
	OS5	0.05	0.2	0.054	0.216	0.056	0.224
	OS6	1.518	6.072	1.522	6.088	1.524	6.096
Conf2	OS1	0.097	0.388	0.092	0.368	0.092	0.368
	OS2	0.137	0.548	0.132	0.528	0.132	0.528
	OS3	0.05	0.2	0.041	0.164	0.041	0.164
	OS4	0.45	1.8	0.441	1.764	0.441	1.764
	OS5	0.082	0.328	0.073	0.292	0.073	0.292
	OS6	1.55	6.2	1.541	6.164	1.541	6.164
Conf3	OS1	0.155	0.62	0.169	0.676	0.164	0.656
	OS2	0.295	1.18	0.309	1.236	0.304	1.216
	OS3	0.137	0.548	0.145	0.58	0.167	0.668
	OS4	0.537	2.148	0.545	2.18	0.567	2.268
	OS5	0.169	0.676	0.177	0.708	0.199	0.796
	OS6	1.637	6.548	1.645	6.58	1.667	6.668

TABLE 3.4

Example of Storage Prices per GB in Dollars

		R1		R2		R3	
		Min	Max	Min	Max	Min	Max
SB	First 1 TB/Month	0.024	0.096	0.0326	0.1304	0.0264	0.1056
	Next 49 TB/Month	0.0236	0.0944	0.032	0.128	0.026	0.104
	Next 450 TB/Month	0.0232	0.0928	0.0315	0.126	0.0255	0.102
	Next 500 TB/Month	0.0228	0.0912	0.0309	0.1236	0.0251	0.1004
SF		0.08	0.32	0.109	0.436	0.088	0.352

respectively. The price interval of each VM is determined according to VM configuration, OS type, and the region located in it. For example, the price interval of VM with Conf1 [1 Cores(s), 0.75 GB RAM, 20 GB Temporary Storage, and Red Hat Enterprise Linux] located in the Southeast Asia is [0.078, 0.312].

Using the selected IaaS services, a case study of eight client's requirement is generated. It consists of four VM, one SB, and three SF as presented in Table 3.6 In addition, data of 60 providers were generated randomly by using

TABLE 3.5

ART for VM and Storage Services per Second

	VM		Storage	
	Min	Max	Min	Max
Conf1	656	734	120	600
Conf2	656	844		
Conf3	370	375		
Conf4	125	135		
Conf5	130	140		
Conf6	650	844		
Conf7	365	375		
Conf8	125	140		
Conf9	350	360		
Conf10	365	370		

TABLE 3.6

Client's Requirements

		IaaS Services			
		VM1	VM2	VM3	VM4
VM	Region	R5	R7	R9	R2
	OS	OS4	OS6	OS2	OS1
	Number of instances	4	3	5	2
	Number of hours	700	700	700	700
	Configuration Name	Conf1	Conf30	Conf40	Conf10
SB	Region	R8			
	Capacity	900			
	Access tier	archive			
	Redundancy	LRS			
		SF1	SF2	SF3	
SF	Region	R6	R4	R1	
	Capacity	800	800	800	
	Redundancy	LRS	LRS	LRS	

the absolute minimum and the absolute maximum. Example of provider's data is presented in Appendix A. The generated data for the three IaaS services for four providers are presented in Tables A.2 and A.3, respectively. For example, the prices per hour for VM with {Conf1, OS} located in R1 from provider 1, provider 2, provider 3, and provider 4 are $0.2431, $0.2089, $0.1539 and $0.3016, respectively, where ART(s) are 703.79/s, 708.65/s, 670.65/s, and 731.48/s, respectively.

3.6   Algorithms' Performance

Theoretically, when the number of generations is increased, the quality of the obtained solutions will be improved. In practice, this rule is not always true because of the stochastic nature of these algorithms. So, the quality of solutions of the three algorithms was checked with different number of generations. Three different generation numbers (100, 150, and 200) were chosen in order to check the effectiveness of changing parameters values on the properties of the algorithms.

A population of 100 solutions was generated randomly and used as an initial population for the three algorithms. The parameters values of each algorithm were chosen to ensure that all algorithms evaluate the same number of solutions. Each algorithm was run 30 times and the analysis was done based on the average weighted objective values of the best solution obtained through all runs. The performance analysis was done in two phases. In the first phase, the performance of each algorithm was checked with the three numbers of generation. While, in the second phase, the performance of the three algorithms was checked with each number of generation. The proposed algorithms was implemented using Visual Basic for Application (VBA) on Excel. All experiments were conducted on machine with Intel Core Duo CPU (2.27 GHz) and 4 GB RAM.

The progress of the best solution for GA, HS, and PSO is shown in Figures 3.14–3.16, respectively. It is noticed that the quality of the solutions of the algorithm was improved when the number of generation increased. For GA and HS, the quality of solutions was improved, but with a little change in the best solution objective value, while PSO continued to improve its solutions with remarkable change.

The best solution progress of the three algorithms with 100 generations is shown in Figure 3.17. In the beginning, it was noticed that the solutions obtained by HS is better than the solutions obtained by GA. However, GA started to improve its solutions after approximately 7,500 solution evaluation and achieve best solution with better objective value than HS. While, the solutions obtained by PSO are better than the solutions obtained by GA and HS.

The box-plot of the three algorithms with 100 generations is shown in Figure 3.18. We can notice that there is a remarkable difference in the median value of the three algorithms. As mentioned before, the algorithms were applied to a minimization problem and based on the presented information from the box-plot, it is notice that PSO has the better properties than GA and HS. This is because PSO has the minimum value and quartile Q0.25 and Q0.75 are closer to the minimum than GA and HS. GA achieve minimum value better than HS, however, HS has better properties than GA as the median

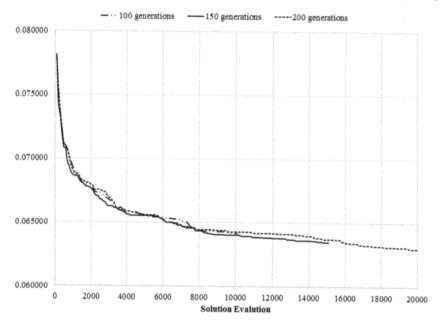

FIGURE 3.14
GA with different generations.

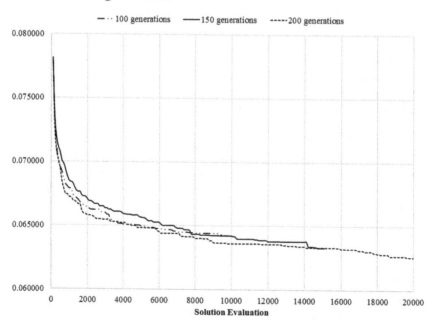

FIGURE 3.15
HS with different generations.

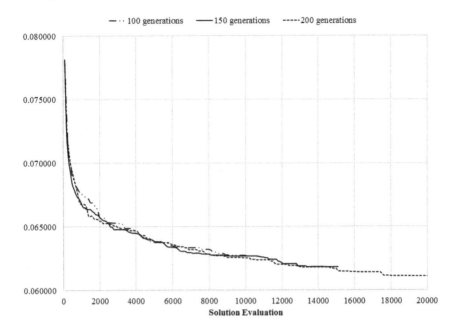

FIGURE 3.16
PSO with different generations.

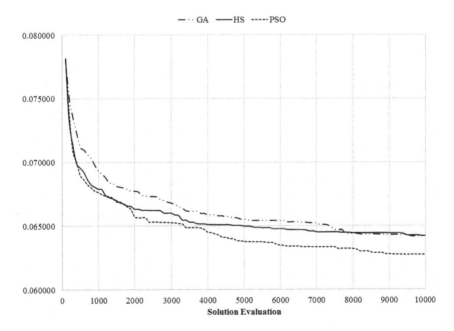

FIGURE 3.17
Best solution progress of the three algorithms with 100 generations.

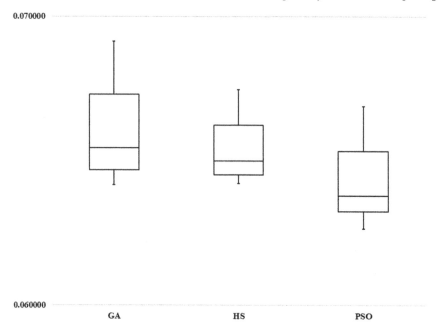

FIGURE 3.18
Box-plot of the three algorithms with 100 generations.

value of HS is smaller than median value of GA and quartile Q0.25 and Q0.75 of HS are closer to the minimum than GA.

The best solution progress of the three algorithms with 150 generations is shown in Figure 3.19. There is no significant difference in the solutions obtained by GA and HS, but HS achieve the minimum value. PSO has remarkable difference in its obtained solutions in comparison with the solutions obtained by GA and HS. From the information of the box-plot of the three algorithms with 150 generations shown in Figure 3.20, we can notice that the minimum value achieved by HS is smaller than the minimum value obtained by GA, however, quartile Q0.25 and Q0.75 of the two algorithms are close. Also, there is a significant difference in the median of PSO; it is smaller than the median of GA and the median of HS and quartile Q0.75 of PSO is smaller than median of GA and HS. Consequently, the solutions acquired by PSO are the best.

Figure 3.21 shows the progress of the three algorithms with 200 generations. The solutions achieved by GA are the worst, while the solutions obtained by PSO are the best. From Figure 3.22, PSO has the minimum value and quartile Q0.25 is close to it, but quartile Q0.50 and quartile Q0.75 are not close. However, PSO has better properties than GA and HS.

From the analysis of the behavior of the three algorithms with different generations, we can conclude that PSO is better than GA and HS; the

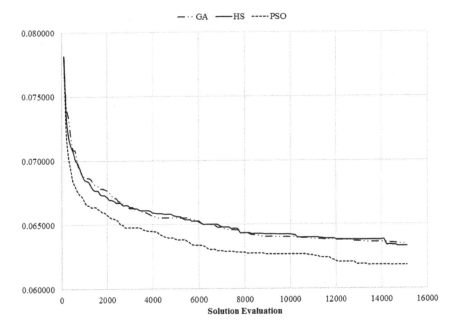

FIGURE 3.19
Best solution progress of the three algorithms with 150 generations.

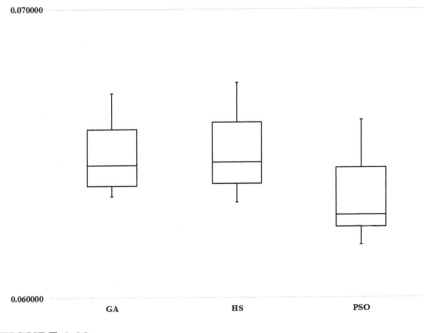

FIGURE 3.20
Box-plot of the three algorithms with 150 generations.

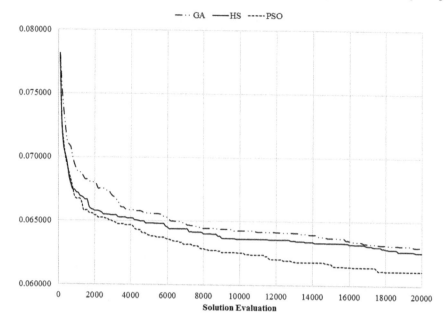

FIGURE 3.21
Best solution progress of the three algorithms with 200 generations.

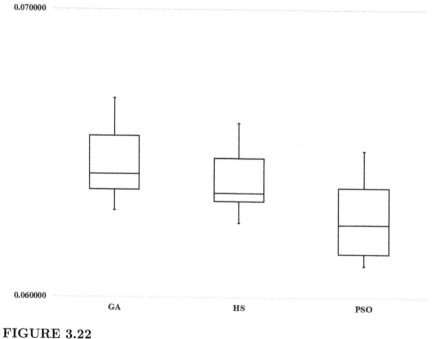

FIGURE 3.22
Box-plot of the three algorithms with 200 generations.

behavior of PSO during different generations is better than GA and HS and it always achieves the minimum value. Besides, the progress of the best solutions obtained by PSO is more robust and stable than the best solutions obtained by GA and HS.

3.7 Conclusion and Future Work

With the diversity of cloud providers, cloud customers faces a challenge of discovering the right providers that can satisfy their requirements. Based on the variation of customer's needs, many research papers proved that dealing with single provider is not sufficient. Thus, this paper presented a general mathematical model to select multiple providers based on minimum cost and maximum performance in order to satisfy customer's requirements and needs. Three heuristics algorithms were proposed to solve the model. Results showed that better solutions obtained by PSO are better than the solutions achieved by GA and HS.

The major limitation of this paper is that the data of the case study were generated randomly and it will be better to test the model on real data. In addition, each of the proposed algorithms has a set of parameters that have a significant influence on the algorithm performance. So, studying the effects of changing values of the different parameters is an important issue. Therefore, further analysis is needed to determine the optimal parameters set.

Appendix A: Example of Case Study Data

TABLE A.1
The Selected Regions

Region ID	Region Name
R1	Southeast Asia
R2	Japan
R3	Australia Southeast
R4	Korea South
R5	West India
R6	West Europe
R7	Germany Central
R8	UK South
R9	US Gov Virginia
R10	Canada Central

TABLE A.2

Random Data for Virtual Machine Service for Four Providers

			\$price/h			ART/s
			R1	**R2**	**R3**	
Provider 1	Conf1					
		OS1	0.2431	0.2132	0.23	
		OS2	0.1341	0.2735	0.4449	
		OS3	0.0463	0.0726	0.0279	
		OS4	1.4586	1.4659	1.1734	703.79
		OS5	0.066	0.2159	0.1696	
		OS6	2.8649	3.2663	2.9	
	Conf2					
		OS1	0.2163	0.2059	0.2887	
		OS2	0.5093	0.2355	0.4429	
		OS3	0.1342	0.1264	0.1534	
		OS4	1.1435	1.0535	0.9086	841.59
		OS5	0.1718	0.1803	0.1071	
		OS6	3.9033	3.3461	2.0374	
	Conf3					
		OS1	0.1894	0.2225	0.3272	
		OS2	0.4649	0.9384	0.7182	
		OS3	0.3022	0.3459	0.4136	
		OS4	1.2521	0.9913	2.0512	375
		OS5	0.5696	0.3348	0.3395	
		OS6	4.5278	5.3705	6.3073	
Provider 2	Conf1					
		OS1	0.2089	0.1358	0.2184	
		OS2	0.4404	0.1793	0.2613	
		OS3	0.043	0.0752	0.0335	
		OS4	0.6191	1.0496	0.853	708.65
		OS5	0.189	0.0838	0.2133	
		OS6	4.3218	1.9642	5.56	
	Conf2					
		OS1	0.2696	0.2271	0.1902	
		OS2	0.3842	0.3491	0.3511	
		OS3	0.1726	0.1296	0.0663	
		OS4	0.7952	1.7607	0.7299	704.2
		OS5	0.3109	0.2794	0.095	
		OS6	3.5079	3.7613	1.9286	
	Conf3					
		OS1	0.5078	0.5459	0.2091	
		OS2	0.9892	0.7704	1.1781	
		OS3	0.2955	0.5508	0.4483	
		OS4	0.9693	1.9268	1.875	373.17
		OS5	0.2552	0.4416	0.4142	
		OS6	4.9329	5.0493	3.2975	

(*Continued*)

TABLE A.2 (*Continued*)
Random Data for Virtual Machine Service for Four Providers

			\$price/h			
			R1	**R2**	**R3**	**ART/s**
Provider 3	Conf1					
		OS1	0.1539	0.1404	0.1563	
		OS2	0.2328	0.2966	0.3003	
		OS3	0.0563	0.058	0.0396	
		OS4	0.7587	0.561	0.9342	670.65
		OS5	0.0596	0.1525	0.2078	
		OS6	4.2423	2.3048	3.0891	
	Conf2					
		OS1	0.1997	0.2548	0.2755	
		OS2	0.448	0.4751	0.5267	
		OS3	0.0576	0.0569	0.0899	
		OS4	1.3746	1.6889	1.0123	658.17
		OS5	0.1206	0.1369	0.2302	
		OS6	3.629	5.1102	5.4653	
	Conf3					
		OS1	0.369	0.5606	0.3381	
		OS2	0.4046	0.9335	0.8498	
		OS3	0.1599	0.2968	0.5594	
		OS4	2.0723	1.3575	1.326	370.5
		OS5	0.2246	0.4883	0.274	
		OS6	5.1745	3.8522	2.1102	
Provider 4	Conf1					
		OS1	0.3016	0.183	0.1639	
		OS2	0.3708	0.2242	0.46	
		OS3	0.058	0.0387	0.0566	
		OS4	1.3302	0.6288	1.3507	731.48
		OS5	0.115	0.0889	0.1712	
		OS6	4.8864	2.9123	1.6205	
	Conf2					
		OS1	0.3054	0.1807	0.3013	
		OS2	0.2797	0.3207	0.4378	
		OS3	0.1734	0.155	0.0919	
		OS4	1.197	0.7468	1.2853	741.58
		OS5	0.1585	0.1658	0.1363	
		OS6	2.2919	4.9028	3.6486	
	Conf3					
		OS1	0.1892	0.2677	0.1886	
		OS2	0.654	1.0577	0.3119	
		OS3	0.3865	0.4987	0.5616	
		OS4	1.9231	1.0817	0.6381	371.85
		OS5	0.4356	0.4418	0.5906	
		OS6	4.3998	5.2755	4.3069	

TABLE A.3

Random Data for Storage Services for Four Providers

		\$ price/GB			
		R1	**R2**	**R3**	**ART/s**
Provider 1					
	First 1 TB/Month	0.0797	0.1232	0.0743	
SB	Next 49 TB/Month	0.0729	0.0445	0.0812	
	Next 450 TB/Month	0.0656	0.0378	0.0636	405.77
	Next 500 TB/Month	0.0507	0.1084	0.0488	
SF		0.2534	0.4053	0.2706	
Provider 2					
	First 1 TB/Month	0.0617	0.1082	0.0979	
SB	Next 49 TB/Month	0.09	0.0354	0.0432	
	Next 450 TB/Month	0.051	0.0724	0.0493	383.39
	Next 500 TB/Month	0.0609	0.088	0.0565	
SF		0.2233	0.336	0.145	
Provider 3					
	First 1 TB/Month	0.0869	0.1135	0.0701	
SB	Next 49 TB/Month	0.0252	0.0706	0.0567	
	Next 450 TB/Month	0.0886	0.0857	0.081	479.39
	Next 500 TB/Month	0.0444	0.0491	0.0431	
SF		0.2201	0.2517	0.0899	
Provider 4					
	First 1 TB/Month	0.0596	0.09	0.0685	
SB	Next 49 TB/Month	0.0712	0.1161	0.0906	
	Next 450 TB/Month	0.0819	0.0586	0.0601	315.25
	Next 500 TB/Month	0.0657	0.0945	0.0718	
SF		0.0924	0.365	0.3103	

Bibliography

[1] Vasilios Andrikopoulos, Steve Strauch, and Frank Leymann. "Decision support for application migration to the cloud." *Proceedings of CLOSER* 13 (2013): 149–155.

[2] Vasilios Andrikopoulos, Alexander Darsow, Dimka Karastoyanova, and Frank Leymann. "CloudDSFthe cloud decision support framework for application migration." *In European Conference on Service-Oriented and Cloud Computing*, pp. 1–16. Springer, Berlin, Heidelberg, 2014.

[3] Ang Li, Xiaowei Yang, Srikanth Kandula, and Ming Zhang. "CloudCmp: comparing public cloud providers." *In Proceedings of the 10th ACM SIGCOMM conference on Internet measurement*, pp. 1–14. ACM, Melbourne, Australia, 2010.

[4] Vasilios Andrikopoulos, Zhe Song, and Frank Leymann. "Supporting the migration of applications to the cloud through a decision support system." *In 2013 IEEE Sixth International Conference on Cloud Computing (CLOUD)*, pp. 565–572. IEEE, Santa Clara, CA, 2013.

[5] Saurabh Kumar Garg, Steve Versteeg, and Rajkumar Buyya. "Smicloud: A framework for comparing and ranking cloud services." *In 2011 Fourth IEEE International Conference on Utility and Cloud Computing (UCC)*, pp. 210–218. IEEE, Victoria, NSW, Australia, 2011.

[6] Ali Khajeh-Hosseini, David Greenwood, James W. Smith, and Ian Sommerville. "The cloud adoption toolkit: supporting cloud adoption decisions in the enterprise." *Software: Practice and Experience* 42, no. 4 (2012): 447–465.

[7] Ali Khajeh-Hosseini, Ian Sommerville, Jurgen Bogaerts, and Pradeep Teregowda. "Decision support tools for cloud migration in the enterprise." arXiv preprint arXiv:1105.0149 (2011).

[8] Michael Menzel, and Rajiv Ranjan. "CloudGenius: decision support for web server cloud migration." *In Proceedings of the 21st international conference on World Wide Web*, pp. 979–988. ACM, Lyon, France, 2012.

[9] Michael Menzel, Marten Schönherr, and Stefan Tai. "(MC2) 2: criteria, requirements and a software prototype for Cloud infrastructure decisions." *Software: Practice and experience* 43, no. 11 (2013): 1283–1297.

[10] Jonas Repschlaeger, Stefan Wind, Ruediger Zarnekow, and Klaus Turowski. "Decision model for selecting a cloud provider: a study of service model decision priorities." (2013).

[11] Adel Alkhalil, Reza Sahandi, and David John. "Migration to cloud computing: a decision process model." *Central European Conference on Information and Intelligent Systems*, Varaždin, Croatia, 2014.

[12] Cloud Service Measurement Index Consortium. http://csmic.org [October, 2018].

[13] Jonas Repschläger, Stefan Wind, Rüdiger Zarnekow, and Klaus Turowski. "Developing a Cloud Provider Selection Model." *In EMISA*, pp. 163–176. Bonn, Germany, 2011.

[14] Narendra Kumar, Shudhi Agarwal, and Babasaheb Bhimrao. "QoS based cloud service provider selection framework." *Research Journal of Recent Sciences* 3 (2014), 7–12.

[15] Songkran Totiya, and Twittie Senivongse. "Framework to support cloud service selection based on service measurement index." *In Proceedings of*

the World Congress on Engineering and Computer Science, vol. 1. San Francisco, 2017.

[16] Gaurav Baranwal, and Deo Prakash Vidyarthi. "A framework for selection of best cloud service provider using ranked voting method." *In 2014 IEEE International conference on Advance Computing Conference (IACC)*, pp. 831–837. IEEE, Gurgaon, India, 2014.

[17] Shruthi Shirur, and Annappa Swamy. "A cloud service measure index framework to evaluate efficient candidate with ranked technology." *International Journal of Science and Research* 4, no. 3 (2015): 1957–1961.

[18] Nitin Upadhyay. "Managing Cloud Service Evaluation and Selection." *Procedia computer science* 122 (2017): 1061–1068.

[19] Yao Wenbin, and Liang Lu. "A selection algorithm of service providers for optimized data placement in multi-cloud storage environment." *In International Conference of Young Computer Scientists, Engineers and Educators*, pp. 81–92. Springer, Berlin, Heidelberg, 2015.

[20] Sai Kiran, Anusha, Gowtham Kumar, and Praveen Kum Rao. "Selection of multi-cloud storage using cost based approach." *International Journal of Computer and Electronics Research* 2, no. 2 (2013): 160–168.

[21] Eiji Oki, Ryoma Kaneko, Nattapong Kitsuwan, Takashi Kurimoto, and Shigeo Urushidani. "Cloud provider selection models for cloud storage services to meet availability requirements." *In 2017 International Conference on Computing, Networking and Communications (ICNC)*, pp. 244–248. IEEE, United States, 2017.

[22] Yashaswi Singh, Farah Kandah, and Weiyi Zhang. "A secured cost-effective multi-cloud storage in cloud computing." *In 2011 IEEE Conference on Computer Communications Workshops (INFOCOM WKSHPS)*, pp. 619–624. IEEE, Shanghai, China, 2011.

[23] Artan Mazrekaj, Isak Shabani, and Besmir Sejdiu. "Pricing schemes in cloud computing: an overview." *International Journal of Advanced Computer Science and Applications* 7, no. 2 (2016): 80–86.

[24] May Al-Roomi, Shaikha Al-Ebrahim, Sabika Buqrais, and Imtiaz Ahmad. "Cloud computing pricing models: a survey." *International Journal of Grid and Distributed Computing* 6, no. 5 (2013): 93–106.

[25] Pandelis G. Ipsilandis, "Spreadsheet modelling for solving combinatorial problems: The vendor selection problem." arXiv preprint arXiv:0809.3574 (2008).

[26] Maysam Eftekhary, Peyman Gholami, Saeed Safari, and Mohammad Shojaee. "Ranking normalization methods for improving the accuracy

of SVM algorithm by DEA method." *Modern Applied Science* 6, no. 10 (2012): 26.

[27] Scott Kirkpatrick, Charles Daniel Gelatt, and Mario Vecchi. "Optimization by simulated annealing." *Science* 220, no. 4598 (1983): 671–680.

[28] Moon-Won Park, and Yeong-Dae Kim. "A systematic procedure for setting parameters in simulated annealing algorithms." *Computers & Operations Research* 25, no. 3 (1998): 207–217.

[29] Russell Eberhart, and James Kennedy. "A new optimizer using particle swarm theory." *In Proceedings of the Sixth International Symposium on Micro Machine and Human Science, 1995. MHS'95*, pp. 39–43. IEEE, Nagoya, Japan, 1995.

[30] James Kennedy, and Russell C. Eberhart. "A discrete binary version of the particle swarm algorithm." In 1997 IEEE International Conference on Systems, Man, and Cybernetics, 1997. Computational Cybernetics and Simulation, vol. 5, pp. 4104–4108. IEEE, Orlando, FL, USA, 1997.

[31] Zong Woo Geem, Joong Hoon Kim, and Gobichettipalayam Vasudevan Loganathan. "A new heuristic optimization algorithm: harmony search." *Simulation* 76, no. 2 (2001): 60–68.

[32] Kang Seok Lee, and Zong Woo Geem. "A new meta-heuristic algorithm for continuous engineering optimization: harmony search theory and practice." *Computer Methods in Applied Mechanics and Engineering* 194, no. 36–38 (2005): 3902–3933.

4

Reliable Data Auditing and ACO-Based Resource Scheduling for Cloud Services

Soumitra Sasmal

Techno Main

Indrajit Pan

RCC Institute of Information Technology

CONTENTS

4.1   Introduction

Virtualization of expert services is in trend over the decades. It helps people to work on different domains even without in-depth knowledge of technology [22]. One of the fastest growing computing paradigms is cloud computing. Cloud computing has gained much attention by empowering people through virtualization of computing infrastructure and services. It can facilitate people or organizations with huge computing infrastructure like CPU core, CPU bandwidth, network infrastructure and peripheral supports. Similarly, it extends support for storage of massive volumes of data and provides a high-end software framework to develop a new application or to simulate and execute some highly resource-centric processes [24].

Figure 4.1 illustrates a classic cloud computing architecture in which the back-end resources combine server, virtual desktop, software platforms, storage or data servers and application software framework [7]. End users are part of the front end and usually connected with the back end via internet through routers and switches. Majority of cloud users can avail the data storage services of cloud infrastructure. Some of the data owners store their data files on cloud storage servers which are later used by many of its associated users [7]. One of the major concerns in maintaining these files in cloud storage is their security. Files stored on a distant virtual storage have a high

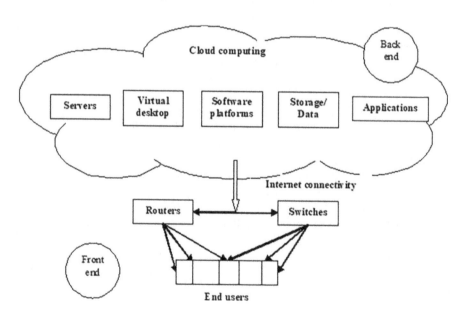

FIGURE 4.1
Cloud computing architecture.

probability of being distorted. Sometimes, one can even upload some content that violates the ethical norms of cloud usage. Thus, auditing of these files is very important both while uploading and during subsequent accesses [12]. In literature, there are many reported works on public auditing of cloud data. A data dynamics-based cloud auditing protocol has been proposed by the work carried out in [12]. The authors have used signature scheme of homomorphic network to perform public auditing and privacy preservation of cloud data. They have implemented an RSA (Rivest–Shamir–Adleman)-based data integration and auditing scheme to execute third-party auditing without revealing any detail of the files content. In another work [15], a challenge–response-based method has been proposed for auditing data files on cloud. Here, the authors have designed two algorithms: the first-phase algorithm uses standard *cryptographic keyGen* algorithm for signature generation which is later used to populate metadata. The second algorithm performs challenge–response-based auditing of the data. A continuous auditing scheme has been proposed for various cloud services through third-party auditor (TPA) in a work reported in [16]. This scheme monitors on-demand access provisioning, competencies of hardware and software components of the servers, and changes in the rule base to maintain reliability and transparency of services. The authors have proposed a secured data storage mechanism here by generating a signature through *MD5 algorithm*. This signature is appended with each encrypted data blocks. *Base* 64 algorithm is proposed for encrypting data blocks which will be uploaded in cloud server. This method also includes a mechanism for checking data integrity by TPA where the auditor can generate a report if any intrusion is detected within the stored data file. A method for encrypting data blocks of each file through advanced encryption standard (AES) algorithm and protecting those encrypted blocks through a message digest generated by *secure hash algorithm 2 (SHA-2)* is proposed in [17]. This is an effective method that claims to be beneficial for third-party auditing, and using this method, auditing can be completed within a fixed amount of time irrespective of the type and size of files.

The works available in literature have each focused on different approaches to minimize the complexity in auditing of data files and the task of resource allocation. Resource allocation is also a very challenging task in cloud computing environment [21,23,25]. This task has a close relation with auditing procedures of data files. Checking the integrity of data files while accessing them is an important process. The process of resource allocation also guides the factors related to file access. It derives the physical location of servers that will be assigned to offer the file access for a particular request. An efficient yet simple resource allocation strategy will ease out verification of data files integrity before granting the access.

There are many research reports available in literature which have worked on different resource allocation strategies in cloud computing environment. A contract-based resource sharing algorithm for geographically distributed and federated cloud infrastructure is proposed in [14]. A scheduling request needs to be submitted to cloud service provider before 24 hours, and based on all requisitions received in a given span of time, this proposed method of [14] performs job scheduling. A load balancing mechanism in virtual machines and resource scheduling is proposed in [18]. This work has taken several simulation setups into consideration to monitor the resource sharing performances by balancing the load of requests through different trials. Swarm optimization approach has been adopted for resource sharing in the work carried out in [19]. They have designed a hybrid method with particle swarm optimization technique and ant colony optimization technique. Different resource units in the system were assumed to be the ants for the method. These ants were primarily initialized with a pheromone value to initiate ant colony optimization approach. Pheromone density on each ant was decided through a particle swarm optimization method. Subsequently, an ant optimization strategy based on pheromone trails was adopted for optimum resource allocation. The authors in [20] have proposed a mechanism for classification of user files to optimize the storage space utilization. They have performed this classification on the basis of usage rate of different files. Mainly the files were categorized into three different clusters (as (i) files which are very frequently used, (ii) files which are averagely used and (iii) files those are rarely used), and different compression techniques were applied on them. Files with heavy usage were given the highest compression. *Huffman encoding* and *Lempel-Ziv algorithms* were used for compression.

Optimization of resource allocation in cloud is always a challenging and most talked about issue. Researchers are rigorously coming up with many proposals and better solutions to address this problem. However, there is lack of in-depth studies on application of swarm-inspired methods at a large extend in optimization of cloud resource scheduling, but application of swarm optimization methods in cloud resource scheduling can be effective, considering the association of multiple factors.

This work will propose a combined dynamic resource allocation and data auditing approach to enhance the reliability of cloud infrastructure in an efficient manner. The work reported in this chapter has utilized ant colony optimization (ACO)-based resource sharing technique along with a novel data auditing scheme. The intended auditing process operates through a safe and sound data upload and verification system. Integrity of the auditing process is ensured through a message digest generated by Merkle tree representation of main data file. Algorithms have been implemented in *Amazon EC2 framework* and compared with the results of some existing research works reported in [12–14] to establish the efficacy of the proposed method.

The next section presents a preliminary discussion on basic ant colony optimization technique and the concept of Merkle tree. Section 4.3 defines the problem, and the proposed method is discussed in Section 4.4. Section 4.5 contains the results and discussion followed by the conclusion in Section 4.6.

4.2   Basic Concepts

4.2.1   Basic Outline of Ant Colony Optimization Method

ACO algorithm is an optimization technique based on the behavior of ants while searching for food or other resources [1]. Ants often move in groups, and they usually try to find out a suitable path between their original location and resource location by a process of evolution. In this biological behavior of ants, they deposit a pheromone along the paths they travel. Initially, they might explore multiple path options to reach their target. Subsequently, a few or more ants start to choose a particular path among the other options. These ants keep on depositing pheromone on the path they travel. Normally, increased amount of pheromone attracts other ants as well. Gradually, the path having dense deposition of pheromone will drag other ants to follow that path. Thus, an optimum path evolves. Dorigo first proposed an optimization technique based on an evolutionary approach in his research work [2]. This evolution was based on the movement behavior of ant colonies. Algorithmic implementation of this method often conceptualizes multiple ants as a multi-agent system where the agents are searching for some optimum results. Mathematical model of ACO suggests that the movement probability (Π_{mn}^a) of such agent (a) from a state m to n depends on two factors. These two factors are the impact of pheromone trail (t_{mn}) upon the connector between m and n, and the logical need of an agent's movement from m to n, which is denoted by (δ_{mn}).

$$t_{mn} = (1 - \rho) \cdot t_{mn} + \Sigma_k \cdot \Delta t_{mn}^k \tag{4.1}$$

This equation represents pheromone update model, where ρ is the evaporation coefficient, $\Sigma_k \cdot \Delta t_{mn}^k$ is the summation of pheromone deposited by all k number of ants on the connection between m and n. Δt_{mn}^k is the amount of pheromone deposited by k^{th} ant, where Equation 4.2 shows the rate of Δt_{mn}^k.

$$\Delta t_{mn}^k = \begin{cases} 0, & \text{if } k^{th} \text{ant not use } m \text{ and } n \text{ connector} \\ \frac{C}{L_k}, & \text{otherwise} \end{cases} \tag{4.2}$$

In the previous equation, C is a constant and L_k is the total travel length of the kth ant.

$$\Pi_{mn}^a = t_{mn}^\alpha \times \delta_{mn}^\beta \tag{4.3}$$

This equation represents the probability (Π) of agent (a) to select the path between m and n. This is governed by the influence (α) of deposited pheromone on path mn (t_{mn}) and the influence (β) of the logical need of adopting the path (δ_{mn}).

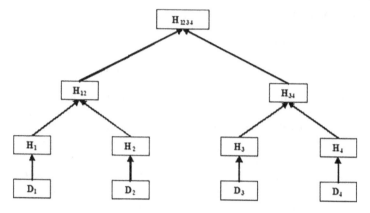

FIGURE 4.2
Merkle tree.

4.2.2   Concept of Merkle Tree

Merkle tree is a hash tree used for cryptographic purposes [3]. It follows the structure of complete binary tree [4]. Every leaf node of Merkle tree contains a data block. Subsequent parent node contains hash of that data contained within the data block. Cryptographic hash function [5] is used to hash these data blocks. A Merkle tree is shown in Figure 4.2. The four leaf nodes are representing four data blocks (D_1, D_2, D_3 and D_4). First, a cryptographic hash function is used to individually hash these data blocks. Penultimate ancestor blocks of the leaf nodes hold the hash of their respective child blocks (e.g. Hash (D_1) is H_1). Afterward, two adjacent hashed blocks are taken together from left to right direction, and they are hashed (e.g. parent blocks of H_1 and H_2 contain their hash H_{12}). Finally, the root contains the final hash block of whole tree. The root with H_{1234} in Figure 4.2 contains Merkle hash of all data blocks.

4.3   Problem Definition

Data auditing within cloud data store is an important task. Optimized dynamic resource allocation to maintain the quality of performance is imperative as well. This proposed work will perform ACO-based dynamic resource allocation and data center auditing to deliver reliable services to cloud users.

Dynamic resource allocation phase has to provide an optimized resource allocation for cloud users. Resources are either infrastructural (CPU and peripheral devices) or service oriented (data center access or application software framework). An optimized solution is needed to allocate these resources among several ongoing tasks depending upon their requisitions. Auditing is associated with data center services. This phase has to deal with the correctness and authenticity of the data stored within the cloud data server to protect cloud users from fraudulent adversaries. Data auditing is mainly performed by a TPA, but the task of TPA becomes challenging under dynamic access allocation. Auditing becomes more crucial to maintain the quality of service for cloud users.

4.4 Proposed Solution

The proposed solution deals with two different tasks. It performs dynamic resource allocation to cloud users and also verifies correctness and authenticity of the data stored in cloud servers. The later part is also known as cloud data auditing. Working principle of the proposed solution along with its primary basics has been discussed in this section.

ACO technique, which is a swarm-based evolutionary optimization process, is applied for efficient and dynamic resource allocation. On the other hand, Merkle tree-based cryptographic hash procedure is used for the purpose of auditing user data. Basic concepts of ACO and cryptographic hash operation based on Merkle tree are briefly described in the following subsections.

4.4.1 Dynamic Resource Scheduling

Resource allocation task primarily performs infrastructure management and service management. Cloud users submit their requests to a task resolver. The task resolver maintains a liaison with a resource manager. The resource manager undertakes the responsibility of resource scheduling [6]. Underlying concept of cloud computing is based on virtualization. The terms coined like task resolver or resource manager are virtual, and these are software based. This framework is illustrated in Figure 4.3. The figure illustrates that a user of cloud services generally interacts with the main task resolver of the cloud through a request submitter module. The task resolver categorizes these submitted requests and forwards them to the resource manager. The resource manager performs the task of scheduling or allocation based on some best practices, and the requisite services are granted to the users subject to satisfying other pre-conditions of availing services [6].

There are multiple resources in the form of CPU core, peripheral units, and some application software frameworks and data centers. The proposed model conceptualizes a system architecture where these units will primarily form an individual sub-cloud network which will be connected with the main cloud. Cloud service provider of the main cloud will control the overall operation. This system model is explained in Figure 4.4.

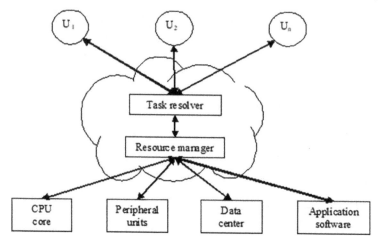

FIGURE 4.3
Schematic diagram of resource management.

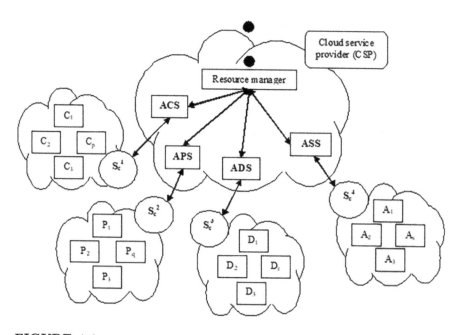

FIGURE 4.4
Proposed resource distribution model.

This figure shows a conceptual view of the proposed resource distribution model. This diagram is mainly focused on the resource management part of the cloud service provider. The proposed work separates four major types of resources (CPU core, peripheral units, data centers and application software frameworks) into four different sub-clouds (namely, S_C^1, S_C^2, S_C^3 and S_C^4). Each of these sub-clouds virtually contains many such resources registered under the same cloud service provider. This work also introduces four sub-schedulers for optimum scheduling of these four types of resources. These four sub-schedulers are ant CPU scheduler (ACS), ant process scheduler (APS), ant data scheduler (ADS) and ant software scheduler (ASS). ACO-based resource scheduling algorithm will work under these respective modules. Decisions of these schedulers are forwarded to resource manager for optimum and dynamic allocation of resources. Working principle of ACO-based resource scheduling along with its pheromone update function is illustrated in the next subsection.

4.4.2 Ant Colony Optimization-Based Resource Scheduling

4.4.2.1 Pheromone Function

Ant resource scheduling method for every resource type (μ) initializes pheromone (t_μ^k) on each instance (k) of the said resource type (μ) upon receiving its requisition by any user (u) through the resource scheduler. Pheromone deposition on every k^{th} instance of a particular resource (μ) is represented by the following equation:

$$t_\mu^k = (1 - \rho) \cdot t_\mu^k + \Sigma_{i-1} \Delta \cdot t_\mu^{i-1} \tag{4.4}$$

$$\rho^k = \frac{x^k}{y_\mu} \tag{4.5}$$

In Equation 4.4, ρ is the standard pheromone evaporation. Evaporation rate for k^{th} ant (ρ^k) is given in Equation 4.5. This is a ratio of the number of allocations (x^k) presently under use for k^{th} instance of the said resource type to the total number of allocations under use for resource type μ, represented by (y_μ). If there are a total i number of $\mu-$ type resources, then $\Delta \cdot t_\mu^{i-1}$ represents pheromone deposited on each of other ($i1$) ants in the present allocation cycle. This deposition value will be 0 if the resource is already allocated to at least one ongoing process and will be ($\frac{\theta}{\omega}$) if it is free. θ is a reputation parameter for a resource instance, which is constant and based on the previous performance feedback of that resource. ω is the total number of idle instances of resource type μ during that allocation cycle.

4.4.2.2 Decision on Resource Scheduling

Decision (Π) on the final selection of kth resource instance (μ^k) will be governed by Equation 4.6. Selection is based on the factor of two parameters. One parameter is related to pheromone trail represented by (($t_\mu^k)^\alpha$), where α is an influence and reputation of that k^{th} instance (($\delta_\mu^k)^\beta$), which is based on its earlier performance feedback. β is the influence of that feedback.

$$\Pi_\mu^k = (t_\mu^k)^\alpha \times (\delta_\mu^k)^\beta \tag{4.6}$$

Overall ACO-based scheduling process is further illustrated in Algorithm 1:

Algorithm 1: ACO-based cloud resource scheduling

A request is received in resource manager of CSP.

The request is forwarded to respective ant resource (μ) scheduler.

The ant resource scheduler connects to sub-cloud of the said resource (S_c^{μ}).

Virtual ant is initialized on all active instances of resource (μ) to compute pheromone following Equation 4.4.

Reputation (δ_{μ}^{k}) of each active instance k is computed.

Selection of resource (μ) instance (k) is performed through equation 4.6.

4.4.3   Auditing of User Data

Major use of cloud computing is associated to its service as huge data centers [7]. A group of users store various data files on cloud data storage for future use. These files are used afterward either by the same users or by a group of other users associated with them. Users of these data files are normally classified into data owners and clients. Data owners store data files on cloud through their authenticated access. Later clients utilize these files for various purposes [8]. As the control and ownership of the files that are transferred to cloud storage is given to a third party, their confidentiality and reliability become major concerns [7]. Confidentiality refers to the concern of data management. Cloud service providers need to ensure that anyone other than the owner or its authorized users wont be able to tamper the files stored in cloud or infringe their contents [9] and only authorized users will get on demand reliable access to these files [10].

The proposed method suggests a data authentication mechanism during any types of data transactions. These transactions may be either upload of user-owned files into the cloud storage servers or subsequent access of those files. This proposed auditing process generates an initial entry-level identity for the files when they are attempted to upload on the server. An owner attempting to upload any file first needs to submit that file to a virtual cloud file manager (CFM). This virtual CFM is an application associated with data upload process. CFM will perform an initial check regarding structural orientation of the file. Then, the file will be split into equal-sized blocks. These blocks will be treated as leaf nodes of Merkle tree (described in Section 4.2) to generate a message digest through Merkle hash. This message digest is unique for every new file. This message digest will be stored under the profile of the owner who has uploaded the file. This method is illustrated in Figure 4.5.

Subsequently, when any user attempts to access that file, another message digest is generated from the then version of the file. This is called user copy of message digest. This user copy of digest and previously saved owners copy of

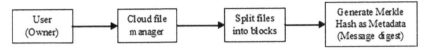

FIGURE 4.5
Audited file upload by owner.

digest will be sent to a TPA to compare and adjudge the integrity. If both these digests found to be the same, then access will be granted to the requesting user. Otherwise, the system will assume that the file has been intruded and corrupted.

After successful integrity check, CFM will grant permission to the user to access the file either in write mode or in read mode. The user will only be able to view the file in read-mode access. On the other hand, the user will be given the permission to modify the file in write-mode access. After gaining write-mode access, the user will become the recent owner of the file and the file will be saved with a new message digest. This new message digest will be tagged with the recent user. Workflow of this access process is illustrated in Figure 4.6.

4.5 Results and Discussion

This proposed system was fully implemented on Amazon EC2 [11] platform. Virtual setup was done on Microsoft Windows Server instance. Initial simulation involved uploading of some data files on the data servers following the process of message digest generation. The proposed Merkle tree method was used to generate this message digest. The other end of the system implemented login by client users in demand of some peripheral and data request. These resources were allocated using ACO-based resource scheduling algorithm. Performance of auditing was measured in terms of following two metrics.

4.5.1 Metadata Generation Time for Data Files of Varying Sizes

In this module, files of different sizes were taken into the process of uploading. System-generated times to compute its metadata at the time of upload were recorded. These times were recorded in milliseconds, and the files sizes are given in kilobytes. Performance details are given in Table 4.1. This table also contains the performance records of the existing works in [12] and [13] on the same parameters. A graphical representation of the observation is given in Figure 4.7.

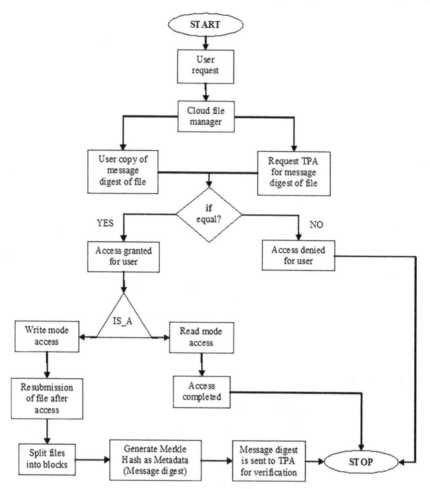

FIGURE 4.6
Auditing of data file access by user.

TABLE 4.1
Time Taken to Generate Metadata Information on Different File Sizes

File Sizes (Kilobytes)	Time Consumed (ms)		
	Proposed Method	[12]	[13]
1,000	90	200	360
5,000	220	360	570
10,000	400	620	1,200
25,000	560	850	1,620
50,000	700	940	1,900
100,000	950	1,150	2,500

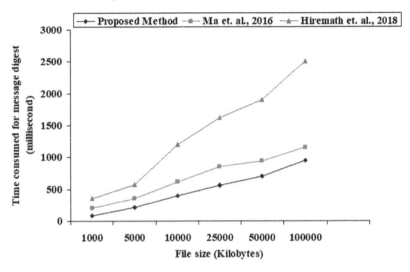

FIGURE 4.7
Comparative performance of the proposed method with [12] and [13] to generate metadata for different file sizes.

4.5.2 Data Verification with Varying Numbers of Tenants

This process records the time taken for data verification when a client requests the access of certain data files. Data verification process checks the integrity of requested data files under different volumes (numbers) of active tenants (clients). Observations are recorded in Table 4.2 along with a comparative performance with the method proposed in [13]. Verification of files integrity measures the average time consumed to take the decision to give access to one requested file. However, files sizes are not taken into consideration while computing this average time. A graphical representation is given in Figure 4.8.

ACO-based dynamic resource allocation task was implemented with the configuration of data centers as depicted in Table 4.3.

TABLE 4.2
Time Taken to Verify File Integrity under Different Volumes of Users

Volume of Users (number)	Average Time Consumed (ms)	
	Proposed Method	**[13]**
5,000	2,250	3,000
10,000	2,675	3,900
25,000	3,210	4,850
30,000	3,540	5,450

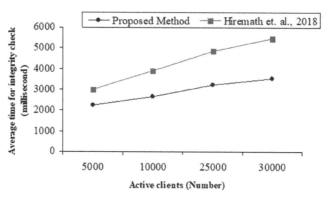

FIGURE 4.8
Comparative performance of the proposed method with [13] on average time consumed for integrity check with different numbers of active clients.

TABLE 4.3
Data Center Configuration

Details	Measurement
Service provider count	15
Number of server per provider	50
Server memory (GB)	16
Server bandwidth (GB)	1
Number of cores per server	12

Implementation of the resource allocation task recorded two different performance metrics.

4.5.3 Successful Resource Allocation with Different Volumes of Active Users

This implementation recorded the percentage (%) of successful resource allocation under different volumes of active users. Observations are recorded in Table 4.4 along with a comparative performance analysis with the work proposed in [14]. Experimental results are also graphically shown on a comparative analysis in Figure 4.9.

4.5.4 Balanced Allocation of Server Resources

According to Table 4.3, the maximum number of available servers is $(500 \times 15) = 7,500$. This section records the average utilization of servers with $25,000$ active users under varying counts of servers. Here the average utilization represents the percentage of servers allocated with at least one job request. The count of servers under different service providers varies in real-time situation.

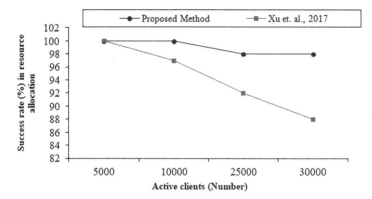

FIGURE 4.9
Comparative on success rate (%) in resource allocation between the proposed method and [14].

Challenge in resource allocation increases when the number of available servers is less due to some reasons. Table 4.5 records a performance metrics on allocation when the count of servers varies due to different reasons. It also contains a performance of work [14] under parallel simulation context. Figure 4.10 shows the comparative analysis.

TABLE 4.4
Success Rate (%) in Resource Allocation with Different Volumes of Clients

	Success Rate (%) of Resource Allocation	
Active Users	**Proposed Method**	**[14]**
5,000	100	100
10,000	100	97
25,000	98	92
30,000	98	88

TABLE 4.5
Average Utilization of Servers under Varying Counts of Active Servers

	Average Utilization (%) of Servers with 25,000 Active Clients	
Active Servers	**Proposed Method**	**[14]**
1,000	90	40
5,000	65	35
10,000	34	35
15,000	22	27
20,000	17	25

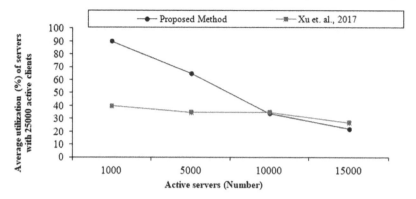

FIGURE 4.10
Average utilization of servers under 25,000 active clients.

4.6 Conclusion

Data auditing and resource allocation are two major activities performed under cloud computing system. Proper functioning of these two tasks is very important for any cloud service provider and cloud framework to sustain in the market as a trusted and reliable service provider. These two activities are closely associated in leveraging the commitment made under service-level agreement (SLA). The present work has proposed a new solution for these two major activities associated with cloud computing. A Merkle tree-based message digest generation technique has been used to facilitate the auditing of the data stored in the cloud storage server during upload and subsequent access operations. Furthermore, an ACO-based dynamic resource allocation scheme has also been proposed for smooth and efficient resource allocation on cloud. These two techniques have been simulated in Microsoft Windows Server virtual instances on Amazon EC2 platform. The proposed auditing technique has been implemented to measure its efficiency in terms of the time taken by the technique to generate message digest during data file upload and the time taken for validation during file access under different volumes of clients. Experimental results show that the performance of the proposed auditing technique is notable and better than a few recent works reported in literature. The ACO-based resource allocation technique has also been implemented to check its proficiency in terms of successful resource allocation and distributing resource requests among available servers. Performance of the proposed ACO-based technique is better than a proposal of one recently published research report. It has been observed that distribution of resource requests among the available servers is wider in the proposed technique. However, a drawback of this distribution may be the power utilization, which has not been explored in this study. Engaging a more number of servers under a lesser load may cause high consumption of power. Still the benefit of wider distribution is reduced failure rate during a high-demand situation when the number of clients and their requests is more. Future work might explore the scope of optimized power utilization under this ACO-based resource allocation scheme. Also a swarm intelligent algorithm-based key generation protocol can be used to implement a public–private key-based digital signature system to protect the privacy and integrity of the uploaded data.

Bibliography

[1] M. Dorigo, V. Maniezzo, and A. Colorni, Ant System: Optimization by a colony of cooperating agents, *IEEE Transactions on Systems, Man, and Cybernetics—Part B*, vol. 26(1), pp. 29–41, 1996.

[2] M. Dorigo, Optimization, Learning and Natural Algorithms, PhD thesis, Politecnico di Milano, Italy, 1992.

[3] R. C. Merkle, A digital signature based on a conventional encryption function, Advances in Cryptology CRYPTO '87, *Lecture Notes in Computer Science*, vol. 293, pp. 369–378, 1988.

[4] S. S. Skiena, *The Algorithm Design Manual*, Springer Science and Business Media, Berlin, 2009. ISBN 978-1-84800-070-4.

[5] S. Al-Kuwari, J. Davenport, and R. Bradford, Cryptographic Hash Functions: Recent design trends and security notions, *In Short Paper Proceedings of Inscrypt 10*, China, Science Press of China, pp. 133–150, 2010.

[6] E. Meriam and N. Tabbane, A survey on cloud computing scheduling algorithms, *Proceedings of 2016 Global Summit on Computer and Information Technology*, Sousse, Tunisia, pp. 42–47, 2016.

[7] J. Spillner, J. Mller, and A. Schill, Creating optimal cloud storage systems, *Future Generation Computer Systems*, vol. 29(4), pp. 1062–1072, 2012.

[8] J. Yanez-Sierra, A. Diaz-Perez, V. Sosa-Sosa, and J. L. Gonzalez, Towards secure and dependable cloud storage based on user defined workflows, *Proceedings of 2015 2nd IEEE Conference on Cyber Security and Cloud Computing*, New York, pp. 405–410, 2015.

[9] H. Xiong, X. Zhang, W. Zhu, and D. Yao, CloudSeal: End to-end content protection in cloud-based storage and delivery services, 96 LNICST, pp. 491–500, 2012.

[10] Outage, Power Outage and Generator Failure Responsible for Instagram, Netflix Blackout, 2012.

[11] https://aws.amazon.com/.

[12] M. Ma, J. Weber, and J. van den Berg, Secure public-auditing cloud storage enabling data dynamics in the standard model, *Proceedings of 2016 Third Internal Conference on Digital Information Processing, Data Mining and Wireless Communications*, Moscow, pp. 170–175, 2016.

[13] S. Hiremath and S. R. Kunte, Ensuring cloud data security using public auditing with privacy preserving, *Proceedings of 2018 3rd International Conference on Communication and Electronics Systems*, Coimbatore, India, pp. 1100–1104, 2018.

[14] J. Xu and B. Palanisamy, Cost-aware resource management for federated clouds using resource sharing contracts, *Proceedings of of 2017 IEEE 10th International Conference on Cloud Computing*, Honolulu, HI, pp. 238–245, 2017.

[15] J. Agarkhed and R. Ashalata, An efficient auditing scheme for data storage security in cloud, *Proceedings of International Conference on Circuits Power and Computing Technologies*, Kollam, India, 2017.

[16] T. Indumathi, N. Aarthy, D. V. Dhanalakshmi, and V. Samyuktha, Third-party auditing for cloud service providers in multicloud environment, *Proceedings of 2017 Third International Conference on Science Technology Engineering and Management*, Chennai, India, pp. 347–352, 2017.

[17] S. Hiremath and S. R. Kunte, A novel data auditing approach to achieve data privacy and data integrity in cloud computing, *Proceedings of 2017 International Conference on Electrical, Electronics, Communication, Computer and Optimization Techniques*, Mysuru, pp. 306–310, 2017.

[18] A. Abdulkarim, I. Muhammed, L. Mohammed, and A. Babayaro, Performance analysis of an improved load balancing algorithm in cloud computing, *American Journal of Networks and Communications*, vol. 8(2), pp. 47–58, 2019.

[19] Z. Yang, M. Liu, J. Xiu, and C. Liu, Study on cloud resource allocation strategy based on particle swarm ant colony optimization algorithm, *Proceedings of IEEE CCIS2012*, Hangzhou, pp. 488–491, 2012.

[20] H. Yahyaoui and S. Moalla, Cloud FC: Files clustering for storage space optimization in clouds, *Proceedings of 2016 IEEE 8th International Conference on Cloud Computing Technology and Science*, Luxembourg, pp. 193–197, 2016.

[21] B. Feng, X. Ma, C. Guo, H. Shi, Z. Fu, and T. Qiu, An efficient protocol with bidirectional verification for storage security in cloud computing, *IEEE Access*, vol. 4, pp. 7899–7911, 2016.

[22] R. Buyya, C. Yeo, S. Venugopal, J. Broberg, and I. Brandic, Cloud computing and emerging IT platforms: Vision, hype, and reality for delivering computing as the 5th utility, *Future Generation Computer Systems*, vol. 25(6), pp. 599–616, 2009.

[23] S. Subashini and V. Kavitha, A survey on security issues in service delivery models of cloud computing, *Journal of Network and Computer Applications*, vol. 34(1), pp. 1–11, 2011.

[24] W. A. Pauley, Cloud provider transparency: An empirical evaluation, *IEEE Security and Privacy*, vol. 8(6), pp. 32–39, 2010.

[25] Z. Birnbaum, B. Liu, A. Dolgikh, Y. Chen, and V. Skormin, Cloud security auditing based on behavioral modeling, *Proceedings of IEEE 9th World Congress on Services*, Santa Clara, CA, pp. 268–273, 2013.

5

TS-GWO: IoT Tasks Scheduling in Cloud Computing Using Grey Wolf Optimizer

Laith Abualigah

Amman Arab University

Mohammad Shehab

Aqaba University of Technology

Mohammad Alshinwan

Amman Arab University

Hamzeh Alabool

Saudi Electronic University

Hayfa Y. Abuaddous

Amman Arab University

Ahmad M. Khasawneh

Amman Arab University

Mofleh Al Diabat

Al Albayt University

CONTENTS

5.1 Introduction

Cloud computing (CC) is the usual prominent solution to deliver large-scale and elastic IT resources (e.g., CPU, Me, and storage) as a service over the internet [2,35]. CC provides a huge pool of IT resources that need to be managed carefully. This is because the performance of CC highly depends on the efficiency of management of cloud service resources [17]. One of the main problems that directly affect the performance of cloud services is tasks scheduling (TS) [15,54]. Decision-making with regard to allocation of the best resources for executing tasks with the consideration of quality of service (QoS) criteria is important not only for meeting cloud users' requirements but also for improving the performance of the whole cloud service. However, TS based on different QoS criteria is described as an NP-hard problem. For example, queries of tens of thousands of end users with different QoS requirements (e.g., performance and cost) need to be distributed across thousands of servers to be served [14].

In the cloud-computing environment, power consumption becomes a critical bottleneck for computer hardware and network infrastructure [19]. Hence, cloud resources require to be designated to save energy and reduce execution time. Consequently, TS and capacity balancing are vital to increasing the performance of cloud service utilizing inadequate resources. The type of TS in CC significantly affects the energy consumption of a cloud data-center. Thus, many researchers have proposed different heuristics [52].

In 2014, a new population-based meta-heuristic algorithm, namely gray wolf optimizer (GWO), was proposed by Mirjalili et al. [36]. GWO imitates the grey wolves' hunting mechanism, which comprises three main steps: starting by tracking the prey, then encircling it, and finally attacking it. This algorithm has shown excellent results in solving several optimization problems in several fields [46]. Many problems related to, for example, data classification [13], feature selection [4], information retrieval [11], dimension reduction [5], data clustering [9], text clustering [6,7], text analysis [10], and benchmark functions [8] can be solved using GWO. There are also other algorithms that can be used to solve these problems [3,48].

In this regard, the arrangement time of tasks that are working on resources is a critical task. Allocating of several tasks with different requirements to be served with a suitable cloud resource is also a daunting and crucial task. This is because resources and tasks in CC are all diverse. Therefore, a multi-objective optimization scheduling model based on gray wolf optimization (GWO) algorithm, called TS-GWO, is proposed. Gray wolf optimization (GWO) method is chosen as it is identical to the combinatorial optimization problem. This method has advantages of enhancing the performance of the TS model and improving the validity of the results obtained. This is due to its capacity to provide high performance even in unknown, challenging search spaces by selecting the optimal solution. Therefore, GWO is applied in the current research to measure the resources and then select the optimal resource for a specific task taking QoS criteria into consideration. The effectiveness of the proposed TS-GWO is assessed based on measurements: Mm values and average cost values.

This chapter contributes to the TS in CC context by:

- Proposing a multi-objective optimization scheduling model based on gray wolf optimization (GWO) algorithm
- Reviewing the more related work in this domain
- Validating the performance of the proposed TS model.

This chapter is organized as follows: Section 5.2 presents the related works. Section 5.3 presents the system model. Section 5.4 presents experimental results and discussions, and finally the conclusion and future work are presented in Section 5.5.

5.2 Related Works

Zuo et al. in [54] introduced a resource cost system for CC by determining the order of jobs on devices with more features. This system builds a connection between the client cost and the demand cost; moreover, client cost demand, and Mm will be the constraints in optimizing problems. Enhanced ant colony algorithm suggested evaluating the performance and the demand cost. To assess the effectiveness of the proposed system, several factors have been considered such as cost, Mm, violation time, and resource utilization. The experimental results demonstrate that the proposed system shows better results compared with other systems.

Reference [15] present an a new version of the particle swarm optimization (PSO) algorithm for TS allocated in CC based on availability, reliability, transportation time, performance time, round-trip time, and transmission cost. The proposed method is called load balancing mutation PSO (LBMPSO). LBMSO has achieved the reliability of the cloud system by using the available resources rule task that fails to designate. The results show a better performance compared with other algorithms. Li, Yibin, et al. suggest a system to reduce the total energy expenditure for smart phones to deal with CC. This method is called dynamic voltage scaling (DVC), which supports power management by reducing the supply voltage and repetition of processors. Li, Yibin, et al. propose an algorithm, called energy-aware dynamic TS (EDTS), to minimize the consumption of energy system for mobile phones. The results show that the proposed work substantially decreases energy expenditure [31].

To reduce the computation cost of CC, the tasks re-allocated are used to the available cloud resources. Reference [50] present a modified PSO (MPSO) algorithm to enhance cloud resource allocation based on price and time. The experiment results show a significant improvement in performance with normal PSO, in terms of utilization, speed, and effectiveness. An algorithm to schedule tasks is proposed in [12] using VM in the cloud environment with reduced time and increased system throughput. This work uses a combination of multi-objective optimization (MOO) and a PSO algorithm called MOPSO for cloud resource allocation. The experimental results show a decrease of 20% and 30% in execution time and waiting time, respectively. Moreover, the algorithm shows an increase of up to 40% in throughput.

In Ref. [47] introduce an algorithm correlates benefit-fairness method depend on weighted-fair queuing system for solving cloud TS problems. The proposed algorithm grouped the optimization problem constraints at a minimum cost and deadline. Then, it applied effective optimization and preference of fairness queue (low, mid, high). The simulation results based on CloudSim simulation show better performance and fairness at the preference level. To overcome the TS challenges in CC, [42] introduce a priority-based scheduling optimization approach. All tasks are grouped based on their priority. To optimize the resource allocation, two constraints have been considered, namely, cost and round-trip time. The results show that the proposed work enhances the allocation operations by decreasing the average round-trip time and cost.

Sreenu and Sreelatha in [49] present a multi-objective algorithm based on whale optimization algorithm (WOA), called W-Scheduler. The fitness value is defined as a multi-objective algorithm by determining the cost of CPU ration and Me. Based on the price and Mm, the WOA will find the optimal scheduling of tasks in the VMs. The proposed algorithm is compared with various algorithms such as PBACO, SPSO-SA, and SLPSO-SA. Experimental results demonstrate that the W-Scheduler shows outstanding performance with a minimum cost and Mm. Based on the ant colony algorithm (ACO), which has a strong convergence rate, [32] propose a genetic algorithm (GA-ACO) for TS in cloud. The simulation results show that the proposed method is considered as the efficient TS method in CC. To optimize the cost and Mm of CC scheduling, a variant of endless PSO algorithm is proposed, named Integer-PSO (IPSO) [18].

Azad and Navimipour in [16] introduce an algorithm based on ant colony and social optimization algorithm, to optimize the energy consumption and the Mm in the cloud environment. The experimental outcomes in the Azure platform show that the introduced algorithm has a significant performance in term of Mm and energy consumption. To solve the multi-QoS problem in cloud scheduling, [22] propose a method with several constraints like security, time-consuming, reliability, and expense in the scheduling process. The technique combines the genetic algorithm (GA) with the ACO algorithm, namely

MQoS-GAAC. The GA algorithm will calculate the initial pheromone values for ACO. Then, ACO is used to find out the optimal resource. Also, to minimize the Mm in TS for CC, [51] propose to use the ACO algorithm. A novel resource scheduling method is proposed in [37] for CC. This work adopts ACO to assign tasks to cloud VMs, namely slave ACO (SACO). The proposed work solves optimization problems by avoiding slave ants that use long ways which may be miscalculated by leaders. Moreover, this work uses diversity and support plans of slave ants. The experimental results show significant performance in cloud server utilization. Based on a small position value rule of the PSO algorithm, [26] introduce an algorithm to reduce the cost of CC processing. The proposed work is compared with the normal PSO, and the experimental results show that the enhanced PSO is more suitable for CC than the normal PSO.

Ramezani and Hussain in [45] developed a multi-objective model based on multi-objective particle swarm optimization (MOPSO) algorithm for optimizing TS and QoS including response time and service cost. To implement and evaluate their model, they extended the Jswarm and Cloudsim packages. In this work, the Jswarm package was extended to multi-objective Jswarm (MO-Jswarm) package by converting the PSO algorithm to MOPSO algorithm. After that, they applied the MO-Jswarm in the Cloudsim toolkit as the task scheduling algorithm. Their simulation results showed that the proposed model significantly increases the QoS. Navimipour and Milani in [40] proposed the cuckoo search algorithm (CSA) to address task scheduling problems in a homogeneous cloud infrastructure. However, the performance of the CSA was not compared with other algorithms such as GA and bee colony. Bitam et al. in [20] proposed the bees' life algorithm (BLA) for cloud scheduling that overtakes GA in terms of execution time. Choudhary and Peddoju in [21] presented a scheduling algorithm that shows a significant improvement in cost and task completion time over sequential assignment. Reference [25] proposed a GA-based load balancing technique for task-level scheduling in Hadoop MapReduce, which outperforms the FIFO algorithm and in terms of delay scheduling and total cost time.

Guo-ning et al. in [24] suggested an optimized algorithm based on genetic simulated annealing algorithm to complete TS in an inefficient way. In their paper [29], Kaur and Sharma presented an improved particle swarm optimization (IPSO) algorithm, which consumes less execution time compared to PSO and simulated annealing algorithms. Due to the rapid growth of mobile devices, various applications have emerged which require high computing power and low latency. Thus, new network architectures have appeared such as fog computing and multi-access edge computing (MEC). Liu et al. in [33] proposed a latency-optimal TS policy for MEC based on Markov decision process (MDP) theory. This policy has achieved a significant reduction in the execution delay compared to the greedy scheduling policy. More related studies can be found in [3,30,41,44,52], and more details are shown in Table 5.1.

TABLE 5.1
Summary of the TS Algorithms

Algorithm	Description	Problem	References
CC	A Introduced a resource cost system for CC by determining the order of jobs on devices with more features	Jobs in cloud computing	Zuo et al. [54]
PSO	Presented a new version of the PSO algorithm for TS and allocation in CC	Allocation in cloud computing	Awad et al. [15]
EDTS	Proposed an algorithm to minimize the energy system energy-aware dynamic TS (EDTS) consumption for mobile phones	Mobile phones	Li et al. [31]
PSO	Pso optimization algorithm for task scheduling on the cloud computing environment	Task scheduling	Tarek et al. [50]
CSSA	Chaotic social spider algorithm for load balance aware task scheduling in cloud computing	Task scheduling	Xavier et al. [52]
GSA	Genetic and static algorithm for task scheduling in cloud computing	Task scheduling	Matos et al. [35]
LBS	Location-based secured task scheduling in cloud	Allocation in cloud computing	Basu et al. [17]
MMS	Multi-objective virtual machine selection for task scheduling in cloud computing	Task scheduling	Naik et al. [39]
OUR	Optimized utilization of resources using improved particle swarm optimization-based task scheduling algorithms in cloud computing	Task scheduling	Kaure et al. [29]
DCT	Delay optimal computation task scheduling for mobile-edge computing systems	Mobile-edge computing systems	Liu et al. [33]

5.2.1 Challenges

Significant difficulties emerge in the CC field when scheduling tasks/jobs to the devices. Currently, there are no methods that can address all critical factors associated with the job scheduling dilemma. The conditions needed for designating the resources raises when the amount of jobs increases. Consequently, optimization methods are not meeting the requirements of cloud cities with a high volume of information. Variety in time-alignment is considered as a complicated process during TS. Most of the current methods solve issues in job scheduling by setting jobs. In unusual circumstances, the expected arrangement was less than the original workloads of the operation, which leads to evil performance of job scheduling based on optimization algorithms. The unexpected complications in evaluation of workloads also affect the remaining parts of the scheduling method [39,53].

5.3 System Model

This area depicts the systemization of the suggested TS component within the cloud ecosystem [27]. In this chapter, we describe a framework that demonstrates assignment planning of CC. The errand supervisor gets the assignments from many users. In this model, the process includes multiple tasks that are arranged and scheduled optimally by the task manager. Each cloud has several VMs, and each one can perform some jobs.

The clients/users yield the assignment demands to the errand director. The assignment chief oversees the database to store each client task. The assignment chief organizes the client errands/tasks and gives the possibility of the assignment to the client. The assignment director contains the data approximate to the state of VMs. The assignment director provides these errands with demands to the errand scheduler. Assignment scheduler could be a gadget that presents the need for approaching chores. The errand scheduler analyzes the (Me) necessity, complications, due date, and the desired resources of the jobs. The cloud environment contains numerous physical devices. The VM display within the electrical device can handle many assignments. The errand scheduler apportions errands to the VMs displayed within the cloud ecosystem.

Consider that there are 100 physical machines in a cloud ecosystem and each electrical device comprises 10 VMs. This could be described as follows:

$$\text{Cloud}, M = \{P_1, P_2, P_3, \ldots, P_100\} \qquad (5.1)$$

where C denotes to the cloud and $P_1, P_2, P_3, \ldots, P_100$ denotes the physical machines displayed within the cloud. The following condition can express the physical machine P_1:

$$\text{Tasks} = \{T_1, T_2, T_3, \ldots, T_i, \ldots, T - 100\} \tag{5.2}$$

where T_1 and T_2 are the first and second errands, respectively. T_i denotes the assignment number i, and $T_1 00$ denotes the 100 jobs. The job supervisor gives the assignments to the controller. Any errand given by the client includes adjustable parameters that are taken, demand, cost demand, due date, and the wanted demand. Consequently, the errand controller prioritizes the errand and plans it in like manner. The prerequisite, complications, due date, and the desired demand of the assignment T_i are analyzed by the scheduler and given to the VM for arranging the tasks [1,35].

As shown in the figure, the term CU_i denotes the take a toll of CPU of the assignment T_i which was characterized by the clients, MU_i denotes the fetched of the Me of assignment T_i characterized by the clients, DU_i denotes the due date for the errand T_i, and BU_i is the demand fetched of T_i.

5.3.1 Multi-objective Design Model

The multi-objective design [54] to solve the TS in the VMs is represented here. The multi-objective design determines the cost estimations of Me and CPU of the group the given VMs provided in the CC ecosystem and determines the demand price. The demand price is determined by calculating the cost value of CPU ration and Me together [39]. Later, the fitness value is determined by combining the demand cost and the Mm value of the arrangement rule.

The fitness function value is determined to evaluate the candidate solutions and choose the optimal one. In the beginning, the cost values of Me and CPU are measured. The subsequent equations are utilized to determine the cost values of CPU and Me in the given VM V_j.

$$C(x) = \sum_{j=1}^{|\text{VM}|} C^{\text{cost}}(j) \tag{5.3}$$

where $C^{\text{cost}}(j)$ denotes the cost value of CPU in the VM number j (V_j), $|\text{VM}|$ denotes the total number of VMs. Next, the $C^{\text{cost}}(j)$ is determined by the following equation:

$$C^{\text{cost}}(j) = C_{\text{base}} \times C_i \times C_{ij} \times C_{\text{Trans}} \tag{5.4}$$

where C_{Trans} denotes the base cost, C_{ij} denotes the CPU of the VM V_j, and C_{ij} denotes the time where the task T_i is preformed in the device R_j. C_{Trans} denotes the transmission cost of the CPU. Here, C_{base} and C_{Trans} are constants (which are the same as in Equations (5.5) and (5.6).

$$C_{\text{base}} = 17/h \tag{5.5}$$

$$C_{\text{Trans}} = 0.0005 \tag{5.6}$$

The cost function value of the Me is determined by the following equation:

$$M(x) = \sum_{j=1}^{|VM|} M^{\text{cost}}(j) \tag{5.7}$$

where $M^{\text{cost}}(j)$ denotes the cost value of the Me of the VM number j V_j, $|VM|$ denotes the total number of the given VMs. Then, $M_{\text{cost}}(j)$ is determined by the following equation:

$$M^{\text{cost}}(j) = M_{\text{base}} \times M_j \times t_{ij} \times M_{\text{Trans}} \tag{5.8}$$

where M_{base} denotes the post cost value of the Me, M_j denotes the Me value of the VM V_j, and t_{ij} denotes the time in which the job T_i is prepared in the resource R_j. M_{Trans} denotes the transportation price value of the Me. The amounts of M_{base} and M_{Trans} are two fixed variables (which are the same as in Equations (5.9) and (5.10)).

$$M_{\text{base}} = 0.05GB/h \tag{5.9}$$

$$M_{\text{Trans}} = 0.5 \tag{5.10}$$

Then, the demand cost value of the agent can be determined using the price need of each VM for both CPU ratio and Me.

$$B(x) = C(x) + M(x) \tag{5.11}$$

where $B(x)$ denotes the demand price value of the agent, $C(x)$ denotes the price value of the CPU, and $M(x)$ denotes the cost value of the Me. Next, the fitness function value is determined as follows (5.12):

$$H(x) = F(x) + B(x) \tag{5.12}$$

where $F(x)$ denotes the Mm function, which must be less than or equal to the finishing of the given task. Mm is determined by the following equation:

$$F(x) =\leq \sum_{i=1}^{|T|} D_i \tag{5.13}$$

where D_i denotes the deadline for the given task T_i . In Equation (5.12), $B(x)$ denotes the demand cost value (it comes from the demand price value) of the tasks that include the CPU ratio and Me cost, and it must be less than or equal to the agent's demand price. The demand price function is determined by the following equation:

$$B(x) =\leq \sum_{i=1}^{|T|} B_i \tag{5.14}$$

where B_i denotes the agent's demand price of given tasks as T_i.

5.3.2   Solution Encoding

The main object of TS is to designate a group of tasks (i.e., 50) to the given VMs (i.e., 10) in accordance with the estimated time of the assignment and the demand cost of the functions. Suppose that there are 50 tasks, and each one is initialized to the conditions from 1 to 10, as shown in Table 5.2. If the first position is 1, then this task 1 is designated to the VM (i.e., V1). If the first position is 8, then the task number 1 is allocated to the (VM) (i.e., V8). Furthermore, all the given tasks are assigned to the given VMs (i.e., V_1V_10).

5.3.3   Gray Wolf Optimizer for Solving the TS

This section explains the proposed method (GWO for solving the tasks scheduling in CC, called TS-GWO) for scheduling tasks using the given number of VMs in the CC ecosystems. The proposed method is introduced with a new multi-objective design model and the GWO. Figure 5.1 presents the description of the proposed TS-GWO for arrangement in cloud ecosystems. The multi-objective design determines the cost use of Me and CPU of all the given VMs and also determines the demand price function through calculating the cost values of Me and CPU [23,28]. The fitness function is determined by calculating both the Mm and the demand price function together. Using the fitness, the designate of the jobs to the given VMs.

TABLE 5.2

Solution Representation for the GWO

T_1	T_2	T_3	...	T_i	...	T_50
V_1	V_3	V_1	...	V_5	...	V_2

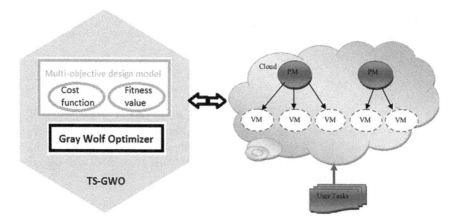

FIGURE 5.1

Description of the proposed TS-GWO for TS problem.

5.3.3.1 Inspiration Source

In 2014, a new population-based metaheuristic algorithm, namely GWO, was proposed by Mirjalili et al. [36]. GWO imitates the grey wolves' hunting mechanism, which comprises three main steps: starting by tracking the prey, then encircling it, and finally attacking it. Alpha (α), beta (β), delta (δ) and omega (ω) refer to the grey wolves' types.

α refers to the leader which is responsible for determining the hunting and sleeping times. Then follows β which is the deputy leader that aids α in making decisions and ensures that the other wolves are doing their tasks. It's worth to mention that β is the strongest candidate to become the leader after the current leader gets old or dies. The third level is δ which serves as connectors between α, β, and ω. In other words, δ monitors ω and sends the information to α and β (i.e., spy). Finally, ω is at the lowest level of the hierarchy, which includes the other wolves except α, β, and δ.

Here, we illustrate and describe the main phases of grey wolf hunting process [34]:

- Approaching the prey
- Pursuing, encircling, and harassing the prey until it stops moving
- Attacking the prey.

5.3.3.2 Mathematical Model and Algorithm

This section shows the mathematical models of GWO phases, for example, social hierarchy, tracking, encircling, and attacking prey. Then follows the pseudo-code of GWO algorithm.

1. **Social hierarchy**
   As mentioned in Section 5.3.3.1, the hierarchical order of gray wolves' types is α, β, δ, and ω. So, the wolves can be mathematically modeled as α for the best solution, β for the second-best solution, δ for the third-best solution, and finally ω for the worst solution.

2. **Encircling the prey**
   As mentioned previously, in hunting, grey wolves chase and encircle the prey. Equations 5.15 and 5.16 present the mathematical model for this phase.

$$\vec{D} = \left| \vec{C} \cdot \vec{X}_p - \vec{X}(t) \right| \tag{5.15}$$

$$\vec{X}(t+1) = \vec{X}_p(t) - \vec{A} \cdot \vec{D} \tag{5.16}$$

   where t and $t+1$ refer to the current and next iterations, respectively. $\vec{A}$ and $\vec{C}$ indicate coefficient vectors (see Equations 5.17 and 5.18), $\vec{X}$ denotes the location vector of a grey wolf, and $\vec{X}_p$ denotes the location vector of the prey.

$$\vec{A} = 2\vec{a} \cdot \vec{r}_1 - \vec{a} \tag{5.17}$$

$$\vec{C} = 2 \cdot \vec{r}_2 \qquad (5.18)$$

where the values of $\vec{a}$ are linearly reduced from 2 to 0 over the course of iterations, as well as r_1 and r_2 are random vectors in $[0, 1]$.

It can be noticed that the location of the grey wolf (X, Y) can be updated based on the prey location (X^*, Y^*). It's worth to mention that setting the values of $\vec{A}$ and $\vec{C}$ vectors can help to achieve a better location. Moreover, the potential updated locations of a grey wolf are shown using 3D space. In this case, vectors r_1 and r_2 allow the grey wolves moving to any location between the points.

3. **Hunting**
   The next step after estimating the prey location is hunting, where the hunting technique of the grey wolf is based on the locations of α, β, and δ, knowing that the prey's location isn't determined exactly. Thus, the values of α, β, and δ are considered as the best solutions. In the case of ω, they'll update their location according to the three obtained solutions. The following equations illustrate the mathematical model of the hunting mechanism:

$$\vec{D}_\alpha = \left| \vec{C}_1 \cdot \vec{X}_\alpha - \vec{X} \right|, \vec{D}_\beta = \left| \vec{C}_2 \cdot \vec{X}_\beta - \vec{X} \right|, \vec{D}_\delta = \left| \vec{C}_3 \cdot \vec{X}_\delta - \vec{X} \right|$$
$$(5.19)$$

$$\vec{X}_1 = \vec{X}_\alpha - \vec{A}_1 \cdot (\vec{D}_\alpha), \vec{X}_2 = \vec{X}_\beta - \vec{A}_2 \cdot (\vec{D}_\beta), \vec{X}_3 = \vec{X}_\delta - \vec{A}_3 \cdot (\vec{D}_\delta)$$
$$(5.20)$$

$$\vec{X}(t+1) = \frac{\vec{X}_1 + \vec{X}_2 + \vec{X}_3}{3} \qquad (5.21)$$

Here, we show the procedures for updating the search locations of the wolves using two-dimensional search space which is based on the locations of α, β, and δ. It can be observed that the final location is determined randomly in a circle using the locations of α, β, and δ.

4. **Attacking the prey (utilization)**
   This phase is also called the exploitation phase. After determining the prey's location, it stops moving. On the other hand, the gray wolves change their locations. The mathematical modal represents this process by decreasing the value of $\vec{A}$ which leads to decrease of $\vec{A}$. We illustrate that $|A| < 1$ allow the wolves to attack across the prey. Therefore, the grey wolves (i.e., ω) change their locations depending on the locations of α, β, and δ. These processes work well, but are prone to stuck in the local optima. So, it is necessary to enhance the exploitation search of GWO.

5. **Searching for prey**
   This subsection presents the mathematical model for increasing the exploration search of GWO algorithm. To achieve this goal, we

should use the value of a with a random value >1 or < -1; in this case, the grey wolves (i.e., ω) were forced to be distant from the prey [38]. We show the inequality $|A| > 1$. In the end, the GWO algorithm is finished by satisfying an end criterion.

Figure 5.2 shows the pseudo-code of the GWO algorithm. The following points describe theoretically how to fix optimization problems:

- Usage of the social hierarchy helps the algorithm to keep the best solutions gained so far over the track of an iteration.

- The surrounding technique determines that a circle-shaped neighborhood includes all solutions that can be expanded to higher dimensions.

- A and C help selected solutions to have hyper-spheres with various random radii.

- The hunting mechanism allows the selected solution to define the probable location of the prey.

- The exploitation and exploration can be controlled by adjusting the values of α and A.

- Using a large value of A helps to improve the exploration search (i.e., $|A| \geq 1$). However, a small value of A is used to enhance the exploitation search (i.e., $|A| < 1$).

- α and C are considered as main parameters of the GWO algorithm.

```
Initialize the grey wolf population Xᵢ (i = 1, 2, ..., n)
Initialize a, A, and C
Calculate the fitness of each search agent
Xₐ=the best search agent
Xᵦ=the second best search agent
Xᵟ=the third best search agent
while (t < Max number of iterations)
    for each search agent
        Update the position of the current search agent by equation (3.7)
    end for
    Update a, A, and C
    Calculate the fitness of all search agents
    Update Xₐ, Xᵦ, and Xᵟ
    t=t+1
end while
return Xₐ
```

FIGURE 5.2
Pseudo-code of the GWO algorithm. [43]

5.4   Experimental Results and Discussions

This section shows the experimental results of the proposed TS in CC using GWO (TS-GWO) for arraigning jobs over the practical devices in the CC ecosystems and the analysis of the introduced algorithm with the other similar approaches, like improved ACO-based scheduling approach (PBACO) [54], self-adaptive learning PSO-based scheduling approach (SLPSO-SA) [55], standard PSO-based scheduling approach (SPSO-SA) [55], and WOA-based scheduling approach (W-Scheduler) [49].

The experimentation of the proposed TS-GWO is implemented in a personal computer (PC) with processor Core i5 and 8GB Me using Windows 10 operating system. The introduced method (TS-GWO) is applied using Java deprogramming with Cloudsim, and the effectiveness of the proposed TS-GWO is assessed using measurements: Mm values and average cost values.

5.4.1   Evaluation Measurements

The evaluation metrics examined for investigating the performance of the proposed TS-GWO method are Mm and cost.

- Mm denotes the total time required for completing all the given assignments. The Mm of the task scheduler methods must be the smallest (minimum Mm).

- The cost denotes the total cost required for programming the tasks to the practical devices (VM).

5.4.2   Results and Discussions

The proposed TS-GWO is assessed based on the testing measurements (Mm and average cost) and also compared with the basic GWO.

Figures 5.3–5.5 show the Mm values using 25 iterations when physical machines = 10 for the population sizes (Pop) of 10, 15, and 20; the Mm values using 50 iterations when virtual machines = 20 for the Pop of 10, 15, and 20; and the Mm values using 50 iterations when physical devices = 30 for the Pop of 10, 15, and 20, respectively.

As shown in Figure 5.3, the numbers of assignments are 100, 200, 300, and 400. In terms of 25 iterations, the proposed TS-GWO got the best Mm results when the tasks are 100 using 20 solutions. When the tasks are 200, the proposed TS-GWO got the best Mm results using 20 solutions. When the tasks are 300, the proposed TS-GWO got the best Mm results using 10 solutions. When the tasks are 400, the proposed TS-GWO got the best Mm results using 20 solutions. From this figure, we conclude that the proposed method (TS-GWO) obtains all the best results when physical machines = 10

THE MAKESPAN

☒ BaisGWO ☒ TS-GWO

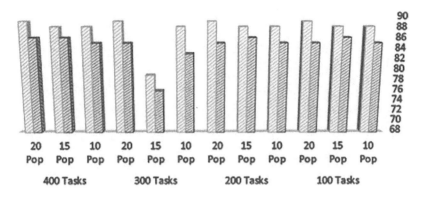

FIGURE 5.3
The Mm values using 25 iterations when physical machines = 10 for the Pop of 10, 15, and 20.

THE MAKESPAN

☒ BaisGWO ☒ TS-GWO

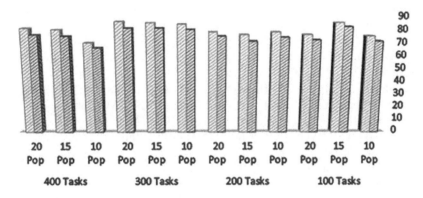

FIGURE 5.4
The Mm values using 50 iterations when physical machines = 20 for the Pop of 10, 15, and 20.

in comparison with other similar purposes, especially when the number of solutions is 20.

As shown in Figure 5.4, the numbers of assignments are 100, 200, 300, and 400. In terms of 50 iterations, the proposed TS-GWO got the best Mm results when the tasks are 100 using 10 solutions. When the tasks are 200,

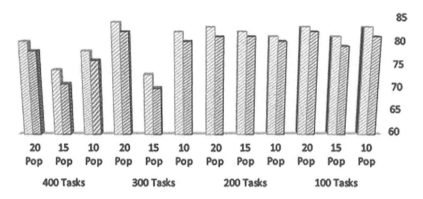

THE MAKESPAN

FIGURE 5.5
The Mm values using 50 iterations when physical machines = 30 for the Pop of 10, 15, and 20.

the proposed TS-GWO got the best Mm results using 15 solutions. When the tasks are 300, the proposed TS-GWO got the best Mm results using 10 solutions. When the tasks are 400, the proposed TS-GWO got the best Mm results using 10 solutions. From this figure, we conclude that the proposed method (TS-GWO) obtains all the best results when physical machines = 30 in comparison with other similar purposes, especially when the number of solutions is 10.

As shown in Figure 5.5, the numbers of assignments are 100, 200, 300, and 400. In terms of 50 iterations, the proposed TS-GWO got the best Mm results when the tasks are 100 using 15 solutions. When the tasks are 200, the proposed TS-GWO got the best Mm results using 10 solutions. When the tasks are 300, the proposed TS-GWO got the best Mm results using 15 solutions. When the tasks are 400, the proposed TS-GWO got the best Mm results using 15 solutions. From this figure, we conclude that the proposed method (TS-GWO) obtains all the best results when physical machines = 30 in comparison with other similar purposes, especially when the number of solutions is 15.

Figures 5.6–5.8 show the demand cost values using 25 iterations when VMs = 10 for the Pop of 10, 15, and 20; the demand cost values using 50 iterations when VMs = 20 for the Pop of 10, 15, and 20; and the demand cost values using 50 iterations when physical devices = 30 for the Pop of 10, 15, and 20, respectively.

As shown in Figure 5.6, the numbers of assignments are 100, 200, 300, and 400. In terms of 25 iterations, the proposed TS-GWO got the best demand cost for all tasks sizes (i.e., 100, 200, 300, and 400). From this figure, we conclude

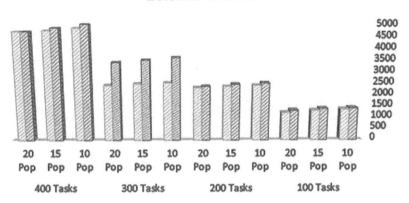

FIGURE 5.6

The demand cost values using 25 iterations when physical machines = 10 for the Pop of 10, 15, and 20.

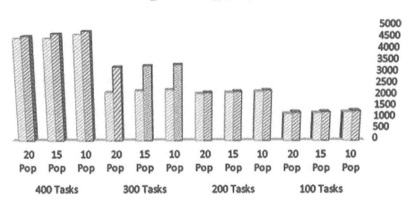

FIGURE 5.7

The demand cost values using 50 iterations when physical machines = 20 for the Pop of 10, 15, and 20.

that the proposed method (TS-GWO) obtains all the best demand costs for all tasks sizes (i.e., 100, 200, 300, and 400) when physical machines = 10 in comparison with other similar methods, especially when the number of solutions is 25.

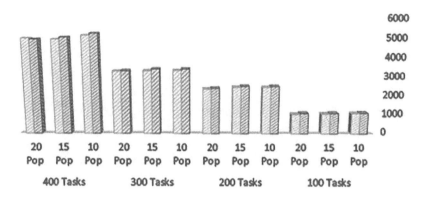

FIGURE 5.8
The demand cost values using 50 iterations when physical machines = 30 for the Pop of 10, 15, and 20.

As shown in Figure 5.7, the numbers of assignments are 100, 200, 300, and 400. In terms of 50 iterations, the proposed TS-GWO got the best demand cost for all tasks sizes (i.e., 100, 200, 300, and 400). From this figure, we concluded that the proposed method (TS-GWO) obtained all the best demand costs for all tasks sizes (i.e., 100, 200, 300, and 400) when physical machines = 30 in comparison with other similar methods, especially when the number of solutions is 50.

As shown in Figure 5.8, the numbers of assignments are 100, 200, 300, and 400. In terms of 25 iterations, the proposed TS-GWO got the best demand cost for all tasks sizes (i.e., 100, 200, 300, and 400). From this figure, we conclude that the proposed method (TS-GWO) obtains all the best demand costs for all tasks sizes (i.e., 100, 200, 300, and 400) when physical machines = 30 in comparison with other similar methods, especially when the number of solutions is 50.

Figures 5.9–5.11 show the average price values using 25 iterations when VMs = 10 for the Pop of 10, 15, and 20; the average cost values using 50 iterations when VMs = 20 for the Pop of 10, 15, and 20; and the average cost values using 50 iterations when physical devices = 30 for the Pop of 10, 15, and 20, respectively.

As shown in Figure 5.9, the numbers of assignments are 100, 200, 300, and 400. In terms of 25 iterations, the proposed TS-GWO got the average cost for all tasks sizes (i.e., 100 tasks when POP = 15, 200 tasks when POP = 15, 300 tasks when POP = 15, and 400 tasks when POP = 20). From this figure, we conclude that the proposed method (TS-GWO) obtains all the average costs

THE AVERAGE COST

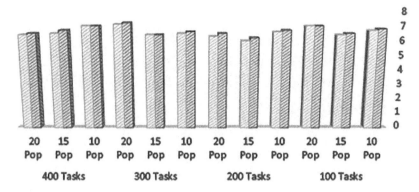

FIGURE 5.9
The average cost values using 25 iterations when physical machines = 10 for the Pop of 10, 15, and 20.

THE AVERAGE COST

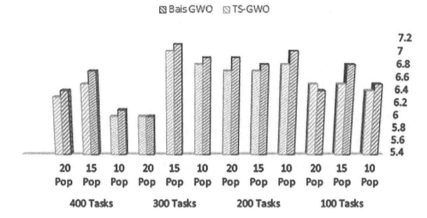

FIGURE 5.10
The average cost values using 50 iterations when physical machines = 20 for the Pop of 10, 15, and 20.

for all tasks sizes when physical machines = 10 in comparison with other similar methods, especially when the number of solutions is 25.

As shown in Figure 5.10, the numbers of assignments are 100, 200, 300, and 400. In terms of 50 iterations, the proposed TS-GWO got the average cost for all tasks sizes (i.e., 100 tasks when POP = 20, 200 tasks when POP = 20, 300 tasks when POP = 20, and 400 tasks when POP = 10). From this figure,

THE AVERAGE COST

⊠ BaisGWO   ⊠ TS-GWO

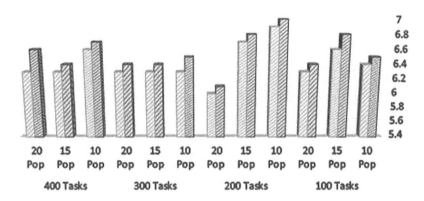

| | 20 Pop | 15 Pop | 10 Pop | 20 Pop | 15 Pop | 10 Pop | 20 Pop | 15 Pop | 10 Pop | 20 Pop | 15 Pop | 10 Pop |

| 400 Tasks | 300 Tasks | 200 Tasks | 100 Tasks |

FIGURE 5.11

The average cost values using 50 iterations when physical machines = 30 for the Pop of 10, 15, and 20.

we conclude that the proposed method (TS-GWO) obtains almost the average costs for all tasks sizes (i.e., 100, 200, and 300) when physical machines = 30 in comparison with other similar methods, especially when the number of solutions is 50.

As shown in Figure 5.11, the numbers of assignments are 100, 200, 300, and 400. In terms of 25 iterations, the proposed TS-GWO got the average cost for all tasks sizes (i.e., 100 tasks when POP = 20, 200 tasks when POP = 20, 300 tasks when POP = 20, and 400 tasks when POP = 20). From this figure, we conclude that the proposed method (TS-GWO) obtains all the average costs for all tasks sizes when physical machines = 30 in comparison with other similar methods, especially when the number of solutions is 50.

5.5   Conclusion and Future Work

This chapter introduces the TS method for scheduling the given jobs or tasks to the available VMs in CC ecosystems. The proposed method is introduced with a new multi-objective design and the GWO algorithm, called TS-GWO. In the beginning, the multi-objective design measures the fitness function by measuring the cost values of the CPU and Me. Next, the demand price value is computed by calculating the price function value of the CPU ratio and Me. Eventually, the fitness function is determined by calculating the Mm value and the demand price value.

The GWO is introduced for finding the optimal scheduling of a group of tasks over the available VMs. The GWO thinks that the currently existing option is optimum to determine the optimal nest option using the best-obtained solution so far. The performance investigation of the proposed scheduling method (TS-GWO) is carried out, and this method is compared with the existing basic GWO for the evaluation analysis using the Mm and cost measure values. According to the empirical outcomes, we conclude that the introduced method can arrange the given jobs to the available VMs optimally, and it produces the minimum Mm values and the minimum average cost values compared with the other.

In future works, we will improve GWO's abilities in effectively finding the best solution by hybridizing it with other algorithm components. Moreover, the proposed algorithm can be implemented to solve other optimization problems.

Bibliography

[1] Mohammed Abdullahi, Md Asri Ngadi, Salihu Idi Dishing, Barroon Isma'eel Ahmad, et al. An efficient symbiotic organisms search algorithm with chaotic optimization strategy for multi-objective task scheduling problems in cloud computing environment *Journal of Network and Computer Applications*, 133:60–74, 2019.

[2] Laith Abualigah and Ali Diabat. A novel hybrid antlion optimization algorithm for multi-objective task scheduling problems in cloud computing environments. *Cluster Computing*, 1–19, 2020.

[3] Laith Abualigah, Mohammad Shehab, Mohammad Alshinwan, and Hamzeh Alabool. Salp swarm algorithm: A comprehensive survey. *Neural Computing and Applications*, 150:1–21, 2019.

[4] Laith Mohammad Abualigah and Ahamad Tajudin Khader. Unsupervised text feature selection technique based on hybrid particle swarm optimization algorithm with genetic operators for the text clustering. *The Journal of Supercomputing*, 73(11):4773–4795, 2017.

[5] Laith Mohammad Abualigah, Ahamad Tajudin Khader, Mohammed Azmi Al-Betar, and Osama Ahmad Alomari. Text feature selection with a robust weight scheme and dynamic dimension reduction to text document clustering. *Expert Systems with Applications*, 84:24–36, 2017.

[6] Laith Mohammad Abualigah, Ahamad Tajudin Khader, and Essam Said Hanandeh. A combination of objective functions and hybrid krill herd algorithm for text document clustering analysis. *Engineering Applications of Artificial Intelligence*, 73:111–125, 2018.

[7] Laith Mohammad Abualigah, Ahamad Tajudin Khader, and Essam Said Hanandeh. Hybrid clustering analysis using improved krill herd algorithm. *Applied Intelligence*, 48(11):4047–4071, 2018.

[8] Laith Mohammad Abualigah, Ahamad Tajudin Khader, and Essam Said Hanandeh. Modified krill herd algorithm for global numerical optimization problems. In *Advances in Nature-Inspired Computing and Applications*, pages 205–221. Springer, 2019.

[9] Laith Mohammad Abualigah, Ahamad Tajudin Khader, Essam Said Hanandeh, and Amir H Gandomi. A novel hybridization strategy for krill herd algorithm applied to clustering techniques. *Applied Soft Computing*, 60:423–435, 2017.

[10] Laith Mohammad Qasim Abualigah. *Feature Selection and Enhanced Krill Herd Algorithm for Text Document Clustering*. Springer, Berlin, 2019.

[11] Laith Mohammad Qasim Abualigah and Essam S Hanandeh. Applying genetic algorithms to information retrieval using vector space model. *International Journal of Computer Science, Engineering and Applications*, 5(1):19, 2015.

[12] Entisar S Alkayal, Nicholas R Jennings, and Maysoon F Abulkhair. Efficient task scheduling multi-objective particle swarm optimization in cloud computing. In *2016 IEEE 41st Conference on Local Computer Networks Workshops (LCN Workshops)*, pages 17–24. IEEE, 2016.

[13] Osama Ahmad Alomari, Ahamad Tajudin Khader, M Azmi Al-Betar, and Laith Mohammad Abualigah. Mrmr ba: A hybrid gene selection algorithm for cancer classification. *Journal of Theoretical and Applied Information Technology*, 95(12):2610–2618, 2017.

[14] AR Arunarani, D Manjula, and Vijayan Sugumaran. Task scheduling techniques in cloud computing: A literature survey. *Future Generation Computer Systems*, 91:407–415, 2019.

[15] AI Awad, NA El-Hefnawy, and HM Abdel_kader. Enhanced particle swarm optimization for task scheduling in cloud computing environments. *Procedia Computer Science*, 65:920–929, 2015.

[16] Poopak Azad and Nima Jafari Navimipour. An energy-aware task scheduling in the cloud computing using a hybrid cultural and ant colony optimization algorithm. *International Journal of Cloud Applications and Computing (IJCAC)*, 7(4):20–40, 2017.

[17] Srijita Basu and Abhishek Anand. Location based secured task scheduling in cloud. In *Information and Communication Technology for Intelligent Systems*, pages 61–69. Springer, 2019.

[18] AS Ajeena Beegom and MS Rajasree. A particle swarm optimization based pareto optimal task scheduling in cloud computing. In *2016 IEEE 41st Conference on Local Computer Networks Workshops*, pages 79–86. Springer, 2014.

[19] Andreas Berl, Erol Gelenbe, Marco Di Girolamo, Giovanni Giuliani, Hermann De Meer, Minh Quan Dang, and Kostas Pentikousis. Energy-efficient cloud computing. *The Computer Journal*, 53(7):1045–1051, 2010.

[20] Salim Bitam. Bees life algorithm for job scheduling in cloud computing. In *Proceedings of The Third International Conference on Communications and Information Technology*, pages 186–191, 2012.

[21] Monika Choudhary and Sateesh Kumar Peddoju. A dynamic optimization algorithm for task scheduling in cloud environment. *International Journal of Engineering Research and Applications (IJERA)*, 2(3):2564–2568, 2012.

[22] Yangyang Dai, Yuansheng Lou, and Xin Lu. A task scheduling algorithm based on genetic algorithm and ant colony optimization algorithm with multi-qos constraints in cloud computing. In *2015 7th International Conference on Intelligent Human-Machine Systems and Cybernetics*, volume 2, pages 428–431. IEEE, 2015.

[23] Tharam Dillon, Chen Wu, and Elizabeth Chang. Cloud computing: Issues and challenges. In *2010 24th IEEE International Conference on Advanced Information Networking and Applications*, pages 27–33. IEEE, 2010.

[24] Guo-ning Gan, Ting-lei Huang, and Shuai Gao. Genetic simulated annealing algorithm for task scheduling based on cloud computing environment. In *2010 International Conference on Intelligent Computing and Integrated Systems*, August 03–05, 2012, Chennai, India. Copyright 2012 ACM 978-1-4503-1196-0/12/08, pages 60–63. IEEE, 2010.

[25] Yujia Ge and Guiyi Wei. Ga-based task scheduler for the cloud computing systems. In *2010 International Conference on Web Information Systems and Mining*, Sanya, China, volume 2, pages 181–186. IEEE, 2010.

[26] Lizheng Guo, Shuguang Zhao, Shigen Shen, and Changyuan Jiang. Task scheduling optimization in cloud computing based on heuristic algorithm. *Journal of Networks*, 7(3):547, 2012.

[27] Soumya Ranjan Jena, Swagatika Tripathy, Tarini P Panigrahy, and Mamata Rath. Comparison of different task scheduling algorithms in cloud computing environment using cloud reports. In *Smart Intelligent Computing and Applications*, pages 33–42. Springer, 2020.

[28] Anthony D Josep, Randy Katz, Andy Konwinski, Lee Gunho, David Patterson, and Ariel Rabkin. A view of cloud computing. *Communications of the ACM*, 53(4):50–58, 2010.

[29] Gunvir Kaur and Er Sugandha Sharma. Optimized utilization of resources using improved particle swarm optimization based task scheduling algorithms in cloud computing. *International Journal of Emerging Technology and Advanced Engineering*, 4(6):110–115, 2014.

[30] AM Senthil Kumar and M Venkatesan. Task scheduling in a cloud computing environment using hgpso algorithm. *Cluster Computing*, 22(1):2179–2185, 2019.

[31] Yibin Li, Min Chen, Wenyun Dai, and Meikang Qiu. Energy optimization with dynamic task scheduling mobile cloud computing. *IEEE Systems Journal*, 11(1):96–105, 2015.

[32] Chun-Yan Liu, Cheng-Ming Zou, and Pei Wu. A task scheduling algorithm based on genetic algorithm and ant colony optimization in cloud computing. In *2014 13th International Symposium on Distributed Computing and Applications to Business, Engineering and Science*, Barcelona, pages 68–72. IEEE, 2014.

[33] Juan Liu, Yuyi Mao, Jun Zhang, and Khaled B Letaief. Delay-optimal computation task scheduling for mobile-edge computing systems. In *2016 IEEE International Symposium on Information Theory (ISIT)*, pages 1451–1455. IEEE, 2016.

[34] Ali Madadi and Mahmood Mohseni Motlagh. Optimal control of dc motor using grey wolf optimizer algorithm. *The Journal of Applied Science, Engineering and Technology* , 4(4):373–379, 2014.

[35] Jocksam G De Matos, Carla K De M Marques, and Carlos HP Liberalino. Genetic and static algorithm for task scheduling in cloud computing. *International Journal of Cloud Computing*, 8(1):1–19, 2019.

[36] Seyedali Mirjalili, Seyed Mohammad Mirjalili, and Andrew Lewis. Grey wolf optimizer. *Advances in Engineering Software*, 69:46–61, 2014.

[37] YoungJu Moon, HeonChang Yu, Joon-Min Gil, and JongBeom Lim. A slave ants based ant colony optimization algorithm for task scheduling in cloud computing environments. *Human-centric Computing and Information Sciences*, 7(1):28, 2017.

[38] Cristian Muro, R Escobedo, L Spector, and RP Coppinger. Wolfpack (canis lupus) hunting strategies emerge from simple rules in computational simulations. *Behavioural Processes*, 88(3):192–197, 2011.

[39] Ketaki Naik, G Meera Gandhi, and SH Patil. Multiobjective virtual machine selection for task scheduling in cloud computing. In *Computational Intelligence: Theories, Applications and Future Directions-Volume I*, pages 319–331. Springer, 2019.

[40] Nima Jafari Navimipour and Farnaz Sharifi Milani. Task scheduling in the cloud computing based on the cuckoo search algorithm. *International Journal of Modeling and Optimization*, 5(1):44, 2015.

[41] Sanjaya K Panda and Prasanta K Jana. An energy-efficient task scheduling algorithm for heterogeneous cloud computing systems. *Cluster Computing*, 22(2):509–527, 2019.

[42] Shachee Parikh and Richa Sinha. Double level priority based optimization algorithm for task scheduling in cloud computing. *International Journal of Computer Applications*, 62(20), 2013.

[43] Ali Parsian, Mehdi Ramezani, and Noradin Ghadimi. A hybrid neural network-gray wolf optimization algorithm for melanoma detection. *Biomedical Research* , 28(8), 2017.

[44] Hua Peng, Wu-Shao Wen, Ming-Lang Tseng, and Ling-Ling Li. Joint optimization method for task scheduling time and energy consumption in mobile cloud computing environment. *Applied Soft Computing*, 80:534–545, 2019.

[45] Fahimeh Ramezani, Jie Lu, and Farookh Hussain. Task scheduling optimization in cloud computing applying multi-objective particle swarm optimization. In *International Conference on Service-Oriented Computing*, pages 237–251. Springer, 2013.

[46] Hasan Rashaideh, Ahmad Sawaie, Mohammed Azmi Al-Betar, Laith Mohammad Abualigah, Mohammed M Al-Laham, M Ra'ed, and Malik Braik. A grey wolf optimizer for text document clustering. *Journal of Intelligent Systems*, 29(1):814830, 2019.

[47] Deepika Saxena, RK Chauhan, and Ramesh Kait. Dynamic fair priority optimization task scheduling algorithm in cloud computing: Concepts and implementations. *International Journal of Computer Network and Information Security*, 8(2):41, 2016.

[48] Mohammad Shehab, Laith Abualigah, Husam Al Hamad, Hamzeh Alabool, Mohammad Alshinwan, and Ahmad M Khasawneh. Moth–flame optimization algorithm: Variants and applications. *Neural Computing and Applications*, 1–26, 2019.

[49] Karnam Sreenu and M Sreelatha. W-scheduler: Whale optimization for task scheduling in cloud computing. *Cluster Computing*, pages 1–12, 2017.

[50] Zahraa Tarek, Magdy Zakria, and Fatma A Omara. Pso optimization algorithm for task scheduling on the cloud computing environment. *International Journal of Computers and Technology*, 13(9), 2014.

[51] Lin Wang and Lihua Ai. Task scheduling policy based on ant colony optimization in cloud computing environment. In *LISS 2012*, pages 953–957. Springer, 2013.

[52] VM Arul Xavier and S Annadurai. Chaotic social spider algorithm for load balance aware task scheduling in cloud computing. *Cluster Computing*, 22(1):287–297, 2019.

[53] Xiaolong Xu, Lingling Cao, and Xinheng Wang. Resource pre-allocation algorithms for low-energy task scheduling of cloud computing. *Journal of Systems Engineering and Electronics*, 27(2):457–469, 2016.

[54] Liyun Zuo, Lei Shu, Shoubin Dong, Chunsheng Zhu, and Takahiro Hara. A multi-objective optimization scheduling method based on the ant colony algorithm in cloud computing. *IEEE Access*, 3:2687–2699, 2015.

[55] Xingquan Zuo, Guoxiang Zhang, and Wei Tan. Self-adaptive learning pso-based deadline constrained task scheduling for hybrid iaas cloud. *IEEE Transactions on Automation Science and Engineering*, 11(2):564–573, 2013.

6

Fact-Checking: Application-Awareness in Data Centre Resource Management

Aaqif Afzaal Abbasi

Foundation University

Almas Abbasi

International Islamic University

Mohammed A. A. Al-qaness

Wuhan University

Ammar Hawbani

University of Science and Technology of China

Ahmed A. Ewees

Damietta University

Yousif A. Alhaj

Wuhan University of Technology

CONTENTS

6.1   Introduction

Computing capabilities in the world of IT are increasing at a rapid pace with constant expansion and growth. The ever-growing IT industry aims to make optimum use of the latest computing technologies to keep pace with emerging and ever-increasing computation demands. In this regard, cloud computing [2,34] is of fundamental importance. It is the art of managing network servers and services on a distributed platform. Cloud computing and related services are often available on requirement basis and can be purchased on a "pay-as-you-go" model. This means that the user does not have to pay extra charges for the use of computing services they do not need and only pay for those they utilize. Similarly, different cloud providing services leverage cloud computing solutions to handle the requirements of resource-intensive IT applications.

The main enabling technology behind the development of cloud computing is virtualization. Virtualization [15,28], is the process of separating the physical computing devices. This results in the creation of "virtual" devices. These devices are later managed to perform some computing tasks. The technique is used to create virtual versions of computing hardware, storage devices, and operating systems. The method was first adopted in the 1960s with an aim to divide a systems resource allocation among various applications [5]. Since then, the term has expanded in use and has now garnered much wider recognition.

System virtualization is the development of a virtual machine (VM). A VM works just like a real computer system [1,21]. The software working on these VMs is separated from the hardware system (see Figure 6.1). Therefore, development of cloud computing is related to the evolution of distributed systems-related disciplines. By using cloud computing, users can benefit from various distributed system-related technologies. The cloud computing paradigm reduces operational costs by helping users to focus only on utilizing IT services instead of wasting it on worrying about the ever-changing IT technologies. The agile, efficient, and cost-effective nature of cloud technology led it to be a major technology of the IT infrastructure.

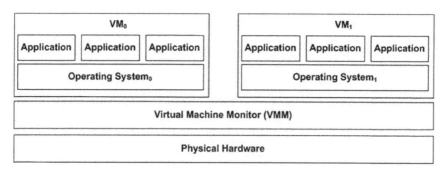

FIGURE 6.1
VMs and operating systems in a virtualized environment

Swarm intelligence (SI) is a concept used in mathematics and computing to derive the collective and decentralized trends in objects and systems. SI often represents the decentralized behavior of objects. The concept relies on artificial intelligence (AI) and was introduced in 1989 by Gerardo Beni and Jing Wang for management of robot technology. SI concepts have often intermingled with various fields of computing and technology to execute resource administration in cloud computing and data centers. The SI concepts consider various communicating elements of cloud computing and data centers as members of a swarm environment. By bringing in intelligence in data centers, the SI concepts play the same role (i.e. improving the resource administration concepts of data center elements).

In general, the idea of SI couples huge networks using loosely coupled technologies, e.g. Internet of Things and cloud computing. SI is a process of bringing intelligence to computing paradigm. In this context, SI can be used to enable the AI functions in data mining and pattern recognition of a particular trend of data hosted on cloud-based systems.

This chapter presents a brief introduction to cloud computing, application-awareness, and related technologies that influence high-performance computing environments.

6.2 Application-Awareness in Cloud Data Centers

Application-awareness [29] is a networking concept that improves a systems capability to keep information about applications to optimize its functionality. The application-aware network makes the best use of available information by using centralized software-defined networking (SDN) features. This enables the network to effectively handle resource allocation challenges in performing various operations [18,25]. Application-aware storage systems often rely on intelligent built-in patterns. If the storage mechanism facilitates application usage conditions, optimized data layouts and caching behaviors can be administered to reach the needed quality of service (QoS) levels.

6.3 Background

Cloud computing [27] technology appeared as one of the important computing prototypes built with the objectives of reducing investments and delivering services on pay-per-usage concepts. Despite attaining success in numerous domains of computing services, cloud computing technologies are still developing. In this regard, there are various workgroups, and industry-standard regulatory authorities are keep trying to address the gaps emerging in the future development and growth. Cloud computing concepts are also vital in helping companies to reduce costs related to infrastructure development and maintenance [8]. Due to the flexible nature of adoption, it allows industries and enterprises to improve their applications to run faster, and also improve their manageability with lower maintenance costs. It also enables cloud management to adjust their resources to meet the ever-fluctuating and un-predictable demands of cloud business models. By using a typical "pay as you go" model, the administrator asks users to choose an appropriate cloud-pricing model. The recent easy availability of high-performance networks and low-cost computing and storage devices has led to the widespread adoption of hardware-depended virtualization function features, cloud-oriented service-centric architectures, and utility computing, which accelerates the explosive development of cloud computing technology.

Cloud computing delivers services through service-oriented architectures, also called service models. In this regard, National Institute of Standards and Technology (NIST) [27] has proposed three primary service models for cloud services, namely Infrastructure as a Service (IaaS), Platform as a Service (PaaS), and Software as a Service (SaaS), as shown in Figure 6.2.

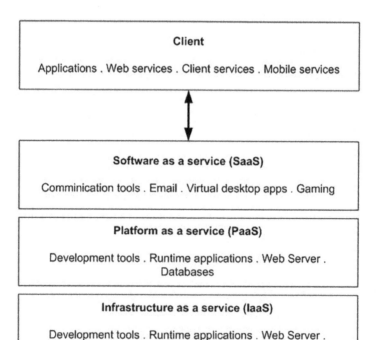

FIGURE 6.2
Cloud service models.

These models are used to deliver abstraction services for cloud functions. According to the Internet Engineering Task Force (IETF), a cloud service is described as a service that offers computing infrastructure for VMs. The same services and resources are provided to subscribers as well. In this regard, IaaS refers only to the services that can be abstracted from the infrastructure details such as the computing machine, its resources, and its location. These services run through VMs by a hypervisor. A large pool of hypervisors residing in a cloud provide support to a large number of VMs. This helps service providers by improving their capacity to scale services as per user needs. The PaaS model offers a platform for development purposes. It also enables application developers to operate their software solutions on a cloud platform without involving users in the complexities of buying and managing an IT infrastructure. In the SaaS service model, users gain system resources access, while cloud service providers manage infrastructure and system applications. The SaaS model is often referred to as "on-demand software". It is also charged on a pay-per-use basis or with a fixed subscription rate.

SDN is a concept in data center networking, where network managers can manage network services by using the abstraction of lower-level functionality. SDN is widely recognized as the future of networking [31]. SDN concept involves the separation of a networks control plane from its data plane. This improves the ease of administration and allows remote access to the data center switches for improved traffic management [32], as shown in Figure 6.3. This feature distinguishes SDN-enabled networks from other

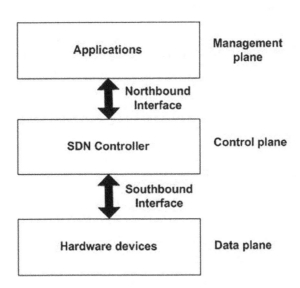

FIGURE 6.3
SDN concepts.

network administration concepts. SDN-enabled networks are vital in dealing with multi-tiered applications as they require definite bandwidth allocation between server instances to assure that user transactions are accomplished within a predefined timeframe under approved service-level agreements (SLAs). The continuation of manual systems for configuration via command-line interfaces (CLIs) has restricted networking from exploring the full features of virtualization. It also increases the operational costs, requires time-consuming network update processes, and is susceptible to introduction of errors [36]. These issues were briefly discussed in [16]. Since their inception, SDN and its first industrial standard OpenFlow [30] have tried to solve many network-related issues such as the network functions virtualization (NFV) [14,17], data centers and cloud networking [26], acceleration in value-added networks, services delivery [3], and network management platforms [9,19].

The software-defined data center (SDDC) concept extends the virtualization concepts to all data center resources. In an SDDC, all functional components of data center, i.e. internetworking, storage, security, and traffic management, are virtualized. According to IT experts, SDDCs will witness a strong market demand in near future [23]. SDDC concept actually comprises a variety of data center concepts that provide, operate and administer data operations using an application program interface (API). The biggest advantage of SDDCs is that cloud service providers need no longer to depend on specialized hardware equipment. All the applications in an SDDC are deployed and run through "logical applications". A large number of vendors, such as VMware, NEC, Microsoft, IBM, Cisco, Juniper, and Hewlett-Packard, are working to develop and establish SDDC approaches. The most commonly cited advantage of SDDCs is the improved efficiency brought about by extending virtualization features throughout the data center, resulting in better administration of network and policy-based services. Several efforts are underway to make the standardization process uniform and to bring it in line with industry standards.

6.4   Research Directions

Application-awareness can be improved in the following major streams.

6.4.1   Application-Aware Data Center Resource Management

Use of SDN to re-engineer data center infrastructure helps to maximize the data center resource utilization. This helps to address issues related to network management systems [22] and network configuration syntax [11], and it often delivers improved network management functions. While the industry is currently witnessing major developments in the areas of network-path optimization [20,24], traffic routing [13,33], latency management [12], and network functions management [35],[1] there is a huge research market for the development of performance fine tuning and related network applications especially designed for data centers. Therefore, the absolute necessity to improve data center infrastructure management has been agreed upon by experts who believe that data center infrastructure management systems are essential for the development of data center efficiency.

6.4.2 Topology of Inference Schemes

Inference schemes use internet exchange providers (IXPs) to provide services and facilities to internet service providers (ISPs). Similar concepts are used for content delivery networks (CDNs) where internet traffic is exchanged between two networks containing featured content. Implementing the SDN concepts through SDDCs at IXPs offers new avenues for solving network domain routing issues. The software-defined exchange (SDX) controller [7,10] is a vital development in this regard as it provides a sequential composition of policies that change inter-domain routing.

6.4.2.1 Managing Cloud Services Admission

Due to the massive amounts of data transfers in cloud computing systems, the huge amount of traffic generated by these services creates serious challenges for cloud infrastructure and service users. Cloud admission control solutions are even more important due to mobile-based applications and services. State-of-the-art research directions regarding admission control of cloud services have been addressed in [4,6]. SDNs are widely used for policy enforcement and isolation to boost cloud service admission in cloud data centers. Services admission control issues in data centers can be resolved through various application-awareness model-based schemes.

6.4.2.2 Intelligent Resource Administration and Resource Optimization

SDNs are used to administer data center resource usage in general. The SI optimization techniques such as ant colony optimization and grey wolf optimization techniques may be used to improve the resource utilization pattern in cloud-based environments.

---

[1]Openflow in mirage. http://www.openmirage.org/blog/announcing-mirage-openflow. (Accessed 1 September, 2016)

6.5   Conclusion

Application-awareness-related challenges in data centers are gaining popularity. However, there is lack of studies in literature on this exciting area of interest in terms of cloud computing. IT and computing professionals over the years have been developing systems that support data center resource consolidation methods. In traditional data centers, traffic flows are assigned with different priorities. This means that the traffic flows dedicated to important applications will get more bandwidth. By doing so, these applications will be processed quickly and thus will spend less time in processing queues. Therefore, the traffic with higher priority will suffer lower data transfer delays. Using this simplistic approach may create issues in resource-hungry applications where several applications compete for bandwidth in the same class. This can also lead to congestion and poor application performance. In addition, this approach only addresses bandwidth allocation issues and misses out latency-related issues.

In view of the above, application-aware networking is a concept that refers to the network-dependent processing of application-layer information. This ensures that the network categorizes the content on the basis of application-layer attributes. This will help network managers to provide optimized application delivery to their clients, guaranteeing application performance, availability, and costs. Applications running in a conventional datacenter sometimes may not require enough bandwidth to justify the occupancy of a switching slot. Therefore, to ensure flexibility, all workloads should be flexible enough to move among servers freely. The same case applies to memory when it migrates outside VM jurisdiction to contact another network service.

In this research study, we try to fill the gaps that are considered as obstacles in adopting a methodical approach for implementing application-awareness concepts in data centers. The recent developments in SDN concepts and their implementation in data centers successfully have paved the way for a new generation of cloud networking paradigm and have also spawned innovation that has achieved key advances in several areas of network resource management. It is evident from these facts that application-awareness in data centers will continue to grow in the future.

Bibliography

[1] Aaqif Afzaal Abbasi and Hai Jin. v-mapper: An application-aware resource consolidation scheme for cloud data centers. *Future Internet*, 10(9):90, 2018.

[2] Aaqif Afzaal Abbasi, Almas Abbasi, Shahaboddin Shamshirband, Anthony Theodore Chronopoulos, Valerio Persico, and Antonio Pescapè. Software-defined cloud computing: A systematic review on latest trends and developments. *IEEE Access*, 7:93294–93314, 2019.

[3] Aaqif Afzaal Abbasi, Hai Jin, and Song Wu. A software-defined cloud resource management framework. In *Asia-Pacific Services Computing Conference*, Bangkok, Thailand, pages 61–75. Springer, 2015.

[4] Narjes Aloulou, Mouna Ayari, Mohamed Faten Zhani, Leila Saidane, and Guy Pujolle. Effective controller placement in controller-based named data networks. In *2017 International Conference on Computing, Networking and Communications (ICNC)*, Silicon Valley, CA, USA, pages 249–254. IEEE, 2017.

[5] Andreas Blenk, Arsany Basta, Martin Reisslein, and Wolfgang Kellerer. Survey on network virtualization hypervisors for software defined networking. *IEEE Communications Surveys & Tutorials*, 18(1):655–685, 2015.

[6] Martin Casado, Teemu Koponen, Scott Shenker, and Amin Tootoonchian. Fabric: A retrospective on evolving SDN. In *Proceedings of the First Workshop on Hot Topics in Software Defined Networks*, Helsinki, Finland, pages 85–90. ACM, 2012.

[7] Qingxia Chen, Fei Richard Yu, Tao Huang, Renchao Xie, Jiang Liu, and Yunjie Liu. An integrated framework for software defined networking, caching, and computing. *IEEE Network*, 31(3):46–55, 2017.

[8] Guilherme Da Cunha Rodrigues, Rodrigo N Calheiros, Vinicius Tavares Guimaraes, Glederson Lessa dos Santos, Marcio Barbosa De Carvalho, Lisandro Zambenedetti Granville, Liane Margarida Rockenbach Tarouco, and Rajkumar Buyya. Monitoring of cloud computing environments: Concepts, solutions, trends, and future directions. In *Proceedings of the 31st Annual ACM Symposium on Applied Computing*, Pisa, Italy, pages 378–383. ACM, 2016.

[9] Anupam Das, Cristian Lumezanu, Yueping Zhang, Vishal Singh, Guofei Jiang, and Curtis Yu. Transparent and flexible network management for big data processing in the cloud. In *Presented as Part of the 5th {USENIX} Workshop on Hot Topics in Cloud Computing*, San Jose, CA, USA, 2013.

[10] Advait Dixit, Fang Hao, Sarit Mukherjee, TV Lakshman, and Ramana Kompella. Towards an elastic distributed sdn controller. *ACM SIGCOMM Computer Communication Review*, 43(4):7–12, 2013.

[11] Nate Foster, Arjun Guha, Mark Reitblatt, Alec Story, Michael J Freedman, Naga Praveen Katta, Christopher Monsanto, Joshua Reich, Jennifer Rexford, Cole Schlesinger, et al. Languages for software-defined networks. *IEEE Communications Magazine*, 51(2):128–134, 2013.

[12] Philip Brighten Godfrey, Matthew Caesar, Ian Haken, Yaron Singer, Scott Shenker, and Ion Stoica. Stabilizing route selection in bgp. *IEEE/ACM Transactions on Networking*, 23(1):282–299, 2014.

[13] Luis Gouveia, Michał Pióro, and Jacek Rak. Preface: Static and dynamic optimization models for network routing problems. *Networks*, 69(1):3–5, 2017.

[14] Hassan Hawilo, Abdallah Shami, Maysam Mirahmadi, and Rasool Asal. NFV: State of the art, challenges and implementation in next generation mobile networks (VEPC). arXiv preprint arXiv:1409.4149, 2014.

[15] Cheol-Ho Hong, Young-Pil Kim, Hyunchan Park, and Chuck Yoo. Synchronization support for parallel applications in virtualized clouds. *The Journal of Supercomputing*, 72(9):3348–3365, 2016.

[16] Fei Hu, Qi Hao, and Ke Bao. A survey on software-defined network and openflow: From concept to implementation. *IEEE Communications Surveys & Tutorials*, 16(4):2181–2206, 2014.

[17] Jinho Hwang, K K Ramakrishnan, and Timothy Wood. NETVM: High performance and flexible networking using virtualization on commodity platforms. *IEEE Transactions on Network and Service Management*, 12(1):34–47, 2015.

[18] Michael Jarschel, Florian Wamser, Thomas Hohn, Thomas Zinner, and Phuoc Tran-Gia. SDN-based application-aware networking on the example of youtube video streaming. In *2013 Second European Workshop on Software Defined Networks*, Berlin, Germany, pages 87–92. IEEE, 2013.

[19] Brendan Jennings and Rolf Stadler. Resource management in clouds: Survey and research challenges. *Journal of Network and Systems Management*, 23(3):567–619, 2015.

[20] Hai Jin, Aaqif Afzaal Abbasi, and Song Wu. Pathfinder: Application-aware distributed path computation in clouds. *International Journal of Parallel Programming*, 45(6):1273–1284, 2017.

[21] Danny Kim. Delivering policy settings with virtualized applications, US Patent 8,078,713, December 13 2011.

[22] Hyojoon Kim and Nick Feamster. Improving network management with software defined networking. *IEEE Communications Magazine*, 51(2):114–119, 2013.

[23] Joseph F Kovar. Software-defined data centers: Should you jump on the bandwagon, 2013.

[24] Bing Li, Maode Ma, and Xinbin Yang. Advanced perceptive forwarding in content-centric networks. *IEEE Access*, 5:4595–4605, 2017.

[25] Gaolei Li, Mianxiong Dong, Kaoru Ota, Jun Wu, Jianhua Li, and Tianpeng Ye. Deep packet inspection based application-aware traffic control for software defined networks. In *2016 IEEE Global Communications Conference (GLOBECOM)*, Washington, D.C., USA, pages 1–6. IEEE, 2016.

[26] Andrew Stephen McGough, Matthew Forshaw, Clive Gerrard, Stuart Wheater, Ben Allen, and Paul Robinson. Comparison of a cost-effective virtual cloud cluster with an existing campus cluster. *Future Generation Computer Systems*, 41:65–78, 2014.

[27] Peter Mell and Tim Grance The nist definition of cloud computing. 2011.

[28] Rashid Mijumbi, Joan Serrat, Juan-Luis Gorricho, Niels Bouten, Filip De Turck, and Raouf Boutaba. Network function virtualization: State-of-the-art and research challenges. *IEEE Communications Surveys & Tutorials*, 18(1):236–262, 2015.

[29] Zafar Ayyub Qazi, Jeongkeun Lee, Tao Jin, Gowtham Bellala, Manfred Arndt, and Guevara Noubir. Application-awareness in sdn. In *ACM SIGCOMM Computer Communication Review*, volume 43, pages 487–488. ACM, 2013.

[30] Charalampos Rotsos, Nadi Sarrar, Steve Uhlig, Rob Sherwood, and Andrew W Moore. OFLOPS: An open framework for openflow switch evaluation. In *International Conference on Passive and Active Network Measurement*, Vienna, Austria, pages 85–95. Springer, 2012.

[31] Scott Shenker, Martin Casado, Teemu Koponen, Nick McKeown, et al. The future of networking, and the past of protocols. *Open Networking Summit*, 20:1–30, 2011.

[32] Seungwon Shin, Vinod Yegneswaran, Phillip Porras, and Guofei Gu. Avant-guard: Scalable and vigilant switch flow management in software-defined networks. In *Proceedings of the 2013 ACM SIGSAC Conference on Computer & Communications Security*, Berlin, Germany, pages 413–424. ACM, 2013.

[33] Sandeep Kumar Singh, Tamal Das, and Admela Jukan. A survey on internet multipath routing and provisioning. *IEEE Communications Surveys & Tutorials*, 17(4):2157–2175, 2015.

[34] Mehdi Sookhak. Dynamic remote data auditing for securing big data storage in cloud computing. PhD thesis, University of Malaya, 2015.

[35] Chenchen Yang, Zhiyong Chen, Bin Xia, and Jiangzhou Wang. When icn meets c-ran for hetnets: an sdn approach. *IEEE Communications Magazine*, 53(11):118–125, 2015.

[36] Soheil Hassas Yeganeh, Amin Tootoonchian, and Yashar Ganjali. On scalability of software-defined networking. *IEEE Communications Magazine*, 51(2):136–141, 2013.

7

Bio-inspired Optimization Algorithms for Multi-objective Task Scheduling in Cloud Computing Systems

A. S. Ajeena Beegom

College of Engineering

M. S. Rajasree

Government Engineering College

CONTENTS

7.1 Introduction

Cloud computing is a recent advancement in the technology field, which allows leasing of computing resources over the Internet for meeting the computing/storage requirements of an individual, small-scale companies, and research organizations. In public cloud platforms, the ownership of the resources resides with an external party, whereas in private cloud systems, it is owned and managed internally within an organization or a university. Benefits of using public cloud systems include reduced infrastructure cost, reduced overhead for the end user, and pay only for the components that are used in the given amount of time. Here scheduling plays an important role, which is the process of mapping the jobs or tasks to the resources available so that the quality of service is guaranteed with a minimum cost of usage.

In a public cloud platform, there are hundreds and thousands of tasks to be scheduled per unit time. Consider the problem of scheduling 20 tasks to 20 virtual machines (VM) considering a single objective function, say minimizing cost. There are $20! = 2,432,902,008,176,640,000$ possible solutions to the problem. Examining all these solutions to find the best solution giving minimum cost is even computationally infeasible. For a cloud computing environment, the variability in task length is also unpredictable. This is because some of the applications that are hosted on the cloud platform may be highly active for a certain number of days of a month or a few hours in a day. Here the pattern of task arrival rate, as well as task length, shows a normal or log-normal distribution trend, but for other days or hours, it shows an entirely random distribution [10]. To model such scenarios, studies on optimal scheduling algorithms are needed since they determine the efficient utilization of resources in the cloud, thereby reducing infrastructure and management cost at cloud service provider's (CSP) end and desirable throughput at the end user's side.

Scheduling large-scale scientific applications to distributed and heterogeneous resources such that certain objective functions such as execution time of tasks (*makespan*) or running and management cost (*cost* in short) are optimized subject to constraints such as storage and processing power, is a well-known problem in the NP-complete category. The problem is a constraint multi-objective optimization problem, where no single solution that is optimal for all the objectives exists. Instead, a trade-off solution can be found, with the property that solutions within this set cannot be improved further in any of the considered objectives without degrading another objective function value. Proposing algorithms and solutions to such problems, especially in the context of cloud computing, is a real challenge to the engineering and scientific community.

For task scheduling in the cloud computing platform, apart from running time or management cost, other considerations like network traffic, user satisfaction in terms of Quality of Service (QoS), and CSP's profit are to be addressed. Many of the existing task scheduling research in cloud computing consider one factor alone as done in [5,6,9,12,14], and tries to find an optimal schedule based on that factor. But single-objective optimization solutions try to optimize one objective function value making another key factor to worse; hence, multiple objective functions characterizing the domain need to be considered. In this work, bi-objective optimization based on *makespan* at the user end and management cost at the CSP's end for task scheduling is considered. The *cost* includes computation cost, communication cost, the overall maintenance cost, and power consumption cost. The proposed solution methodology can be applied to any number of objective functions. We have focused on the scheduling of a large set of parallel independent tasks of different sizes, as they form the bottleneck in many of the scientific applications. The weighted sum approach for multi-objective optimization is used, and a comparative study on the performance of nature-inspired optimization algorithms such as Particle Swarm Optimization (PSO) algorithm, Artificial Bee Colony (ABC) algorithm, Genetic Algorithm (GA), and Ant Colony Optimization (ACO) algorithm is done. These algorithms are applied to the same sets of task data and are compared against Greedy Randomized Adaptive Search Procedure (GRASP), which is a list-based meta-heuristic search technique.

7.2   Related Work

A comprehensive study on the evolutionary computation approaches for resource scheduling in the cloud computing systems is presented by Zhan et al. [18]. A taxonomy of various approaches in the literature for task scheduling using GA, PSO, and ACO in the SaaS layer is presented in detail in this work. Since our work falls into *scheduling for negotiations* category, the literature review is focused on this type of work alone. Our previous work [2] uses PSO technique for bi-objective task scheduling in the cloud environment using weighted sum approach and proposed Integer-PSO technique. Here a method to find discrete values after the velocity and position update of *particles* is proposed. M. Feng et al. [7] use the PSO algorithm to solve resource allocation in cloud computing. They have considered total execution time, resource reservation, and QoS of each task as optimization objectives and used the pareto dominance principle to find optimal solutions. L Guo et al. [8] address the task assignment problem in cloud computing considering makespan and cost using PSO algorithm. E.S. Alkayal [1] applies the PSO technique using a ranking strategy for minimizing waiting time and maximizing system throughput.

A hyper-heuristic approach to scheduling in the cloud is proposed in [16], which uses an integration of different heuristics algorithms such as simulated annealing, GA, PSO algorithm, and ACO algorithm, and in each iteration of the algorithm, any one of these techniques is selected for finding the schedule. MOEA/D technique is used by X. Wang et al. [17] for bi-level multi-objective task scheduling which uses map-reduce framework. G.F. Elhady and M.A. Tawfeek [6] have done a comparison between ACO, PSO, and ABC algorithms for cloud task scheduling considering a single objective function. Application of the GA framework for task scheduling in cloud computing systems is addressed in our prior work [3]. Bi-objective optimization using NSGA-II for optimizing *makespan* and cost for hybrid cloud platform is done by V.A. Leena et al. [11].

7.2.1   Contribution of Chapter

1. Analysis of the suitability of bio-inspired optimization algorithms such as ABC algorithm, PSO algorithm, GA, and ACO algorithm for the cloud environment.

2. Study on multi-objective optimization for cloud task scheduling for parallel independent tasks considering *makespan* at the user's end and *cost* at the CSP's end, two conflicting objective functions.

3. Experimental analysis based on different data sets generated using an established procedure to characterize normal traffic and bursty traffic in the cloud environment.

7.3   Mathematical Formulation

For a better understanding of the mathematical model, the variables used are presented in Table 7.1 along with a brief description.

Assuming that an application consists of N tasks which are independent, then n out of N are scheduled to m VMs in each time window, where the value of n is limited by m. To completely execute all tasks in the application, $k = \frac{N}{n}$ similar epochs are needed. For each type of VM instances say *small, medium, large, etc.*, cost of usage and the processing power are different. Let P_j represent the processing power of jth VM instance type, where j ranges from 1 to m and E_j represents its execution cost for unit time. Given the task length T needed to execute each task in a *'small'*-type VM, the optimization objectives for N tasks are as follows:

$$\text{Minimize} \quad M = \sum_{p=1}^{k} \sum_{i=1,j=1}^{n,m} T_{pi} * P_j * z_{ij} \qquad (7.1)$$

$$\text{Minimize} \quad C = \sum_{p=1}^{k} \sum_{i=1,j=1}^{n,m} E_j * T_{pi} * P_j * z_{ij} \qquad (7.2)$$

TABLE 7.1

Variables and Parameters

Variables	Description	Variables	Description
N	Total number of tasks	n	No. of tasks in an epoch
m	No. of virtual machines	k	No. of epochs
P	Processing power	E	Execution cost
T	Estimated task length	z	Decision variable
M	*Makespan* function	C	*Cost* function
Z	Objective function	θ	Weight parameter

with constraints

$$z_{ij} = \begin{cases} 1 & \text{if TASK}_i \quad \text{scheduled to} \quad VM_j \\ 0 & \text{otherwise} \end{cases} \tag{7.3}$$

$$n \le m \tag{7.4}$$

and

$$\sum_{i=1}^{n} z_{ij} = m \tag{7.5}$$

Equation (7.3) ensures that any task in the task list is assigned to some VM, and Equation (7.4) for decision variable z_{ij} ensures that there exist a sufficient number of VMs to schedule each task. Constraint function (7.5) ensures that all tasks in the task list are scheduled.

7.3.1 Multi-objective Optimization

A multi-objective optimization problem consists of optimizing a vector of size n where the objective function $F(x) = (f_1(x), f_2(x), \ldots, f_n(x))$. Classical approaches to solving a multi-objective optimization problem include combining multiple objectives into a single objective function, by assigning weights to different objective functions [13,15]. The advantage is that they generate a single solution to the problem based on the assigned weights, and this approach is easy to solve. But the relative weight among objective functions needs to be known in advance. To find an optimal task schedule considering the two objective functions M (Equation (7.1)) and C (Equation (7.2)) simultaneously using the weighted sum approach, the optimization problem is represented using the following formula:

$$\text{Minimize} \quad Z = \theta * C + (1 - \theta) * M \tag{7.6}$$

where θ represents the relative preference of one objective function over the other in the range $[0, 1]$. When $\theta = 0$, the optimization problem becomes that of minimizing M alone, and when $\theta = 1$, the problem becomes minimizing C alone, producing *makespan-optimal* task schedule and *cost-optimal* task schedule, respectively.

7.4 Solution Methodologies

The different optimization techniques used in this study and the corresponding algorithm to solve the task scheduling problem are explained in this section.

7.4.1   Particle Swarm Optimization Technique

PSO algorithm is one of the population-based stochastic optimization techniques. Initially, the PSO algorithm generates randomly a set of N solutions called *particles* in the d-dimensional search space. Each *particle* is represented by a D-dimensional vector X_i which stands for its position $(x_{i1}, x_{i2}, \ldots, x_{id})$ in space. Each *particle* maintains its position and velocity. It also maintains the best fitness value it has achieved so far during the search and the position that achieved this fitness (pbest). The PSO algorithm also maintains the best fitness value achieved among all *particles* in the swarm and the corresponding position (gbest). This algorithm was proposed initially for solving problems in the continuous domain. Since the task scheduling problem is in the discrete domain, Integer-PSO technique [2] is used in this work. The velocity (v) update, constrained by $v_{\min}$ and $v_{\max}$, and position (x) update mechanism of each *particle* are given in Equations (7.7)–(7.10).

$$v_{id}^{n+1} = w * v_{id}^{n} + c_1 * r_1 * (\text{pbest}_{id}^{n} - x_{id}^{n}) + c_2 * r_2 * (\text{gbest}_{d}^{n} - x_{id}^{n}) \quad (7.7)$$

$$Y_{id}^{n+1} = ceil((x_{id}^{n} + v_{id}^{n+1}) * \beta) \text{ where } \beta = 10^y \quad (7.8)$$

$$\text{Pos}_{id}^{n+1} = (Y_{id}^{n+1}) \quad modulo \quad m \quad (7.9)$$

$$xc_{id}^{n+1} = \begin{cases} \text{Pos}_{id}^{n+1} & \text{if } \text{Pos}_{id}^{n+1} > 0 \\ m & \text{otherwise} \end{cases} \quad (7.10)$$

where $i = 1, 2, \ldots, N$, $n = 1, 2, \ldots$, max, w is the inertia weight, c_1 and c_2 are acceleration coefficients, and r_1 and r_2 are two uniformly distributed random numbers in the interval $[0, 1]$. New position x_{id}^{n+1} is obtained from xc_{id}^{n+1} by removing duplicate entries, if any, in the assignment generating (Task, VM) schedule for the next iteration.

PSO-Based Task Scheduling Algorithm is given as Algorithm 1. Permutation encoding is used to represent each population. For example, (5, 4, 1, 3, 2) represents a (Task, VM) assignment schedule showing Task$_1$ assigned to VM_5, Task$_2$ to VM_4, etc. In step 3–6 of the algorithm, N initial populations (schedules) are generated randomly and are checked to see that all the constraints are satisfied. The velocity of all *particles* is also initialized. In steps 9 and 10 depending on task length T_i, processing power P_j of VM_j and cost E_j of VM_j, the values of M (Equations (7.1)) and C (Equation (7.2)) are calculated. After appropriate scaling of these values, the value of Z (equation (7.6)) of each of these populations is also calculated for the given θ value. From this set, pbest and gbest are assigned (steps 11–16).

Steps 17 and 18 of the algorithm generates new schedules from the current population using the velocity and position update rules (Equations (7.7)–(7.10)). If (5, 4, 1, 3, 2) is one *particle* with pbest as (1, 3, 4, 2, 5) and gbest as (1, 3, 2, 4, 5) and if $c_1 = c_2 = 0.2$ and $w = 0.6$, $r_1 = 0.5$, $r_2 = 0.3$ and current velocity $v_{id}^{n} = 1.4$, then $v_{id}^{n+1} = (0.2, 0.68, 1.2, 0.8, 1.32)$ and

---

Algorithm 1 PSO-Based Task Scheduling Algorithm

1: Initialize N, max, w, c_1, c_2, v_{min}, v_{max}, θ, $High$
2: $n = 1$, $i = 1$, Zgbest $= High$, pbest$(n)(i) = null$, $gbest = null$
3: **for** $(i = 1$ to N$)$ **do**
4:     $P(n)(i) \leftarrow$ Generate random schedule
5:     Initialize $v(n)(i)$ randomly in the range $[v_{min}, v_{max}]$
6: **end for**
7: **while** $(n <=$max$)$ **do**
8:     **for** (each $P(n)(i), i \leq N$) **do**
9:         Calculate $M(i)$ and $C(i)$ as per Equations (7.1)–(7.2)
10:         $Z(i) = \theta * C(i) + (1 - \theta) * M(i)$
11:         **if** $(Z$pbest$(i) > Z(i))$ **then**
12:             pbest$(i) = P(n)(i)$, Zpbest$(i) = Z(i)$
13:         **end if**
14:         **if** $(Z(i) < Z$gbest$)$ **then**
15:             gbest$=$ pbest(i) , Zgbest $= Z(i)$
16:         **end if**
17:         $v(n + 1)(i) \leftarrow$Update $v(n)(i)$ as per Equation (7.7)
18:         $xc(n + 1)(i) \leftarrow$Update $xc(n)(i)$ as per Equations (7.8)–(7.10)
19:         $x(n + 1)(i0 \leftarrow$Remove duplicate from $xc(n + 1)(i)$
20:         $P(n + 1)(i) = x(n + 1)(i)$, $i = i + 1$
21:     **end for**
22:     $i = 1$, $n = n + 1$
23: **end while**
24: Output gbest

$Y_{id}^{n+1} = (5.2, 4.68, 2.2, 3.8, 3.32)$. If $y = 2$, then $xc_{id}^{n+1} = (5, 3, 5, 5, 2)$ which is changed to (5, 3, 1, 4, 2) on duplicate removal to ensure that every VM is assigned a task (step 19). Other constraints are also checked for each *particle*, and if there are any violations on constraints, that *particle* is replaced with its previous iteration value else new position is assigned (step 20). Steps 7–23 are repeated until max iteration is reached. On completion, gbest is returned, as it gives the best schedule achieved among all the *particles* in the population in all generations.

7.4.2   Artificial Bee Colony Algorithm

ABC algorithm is another population-based meta-heuristics algorithm that simulates the foraging behavior of honey bees in nature. A honey bee swarm consists of food sources, employed bees, onlooker bees and scout bees. Each food source represents a solution to the optimization problem to be solved and have a *fitness* value representing its nectar amount. The solutions are subjected to repeated update through employed bees, onlooker bees and scouts for a predetermined number of cycles. A candidate food position v_{ij} is computed from the current food position x_{ij} by employed bees using the following formula:

$$v_{ij} = x_{ij} + \phi_{ij} * (x_{ij} - x_{kj}) \qquad (7.11)$$

where k, j and ϕ are three random numbers in the range $[1, E]$, $[1, U]$ and $[-1, 1]$, respectively, with E as the number of employed bees. To get v_{ij} values as discrete quantities with m as the number of available VMs, the following equations are applied:

$$Y_{ij} = (\text{round}(|v_{ij}|)) \text{ modulo } m \qquad (7.12)$$

$$v_{ij} = \begin{cases} Y_{ij} & \text{if } Y_{ij} > 0 \\ m & \text{otherwise} \end{cases} \qquad (7.13)$$

Fitness value of the new solution is computed, and if it is better than the current food position, the employed bee replaces the current food position with the new position else retains the current food position. When all employed bees complete the update process, they share this information with onlooker bees (*waggle dance*). An onlooker bee chooses a food source probabilistically using equation (7.14), where N denotes the total number of bees and f their fitness.

$$p_i = \frac{f_i}{\sum_{n=1}^{N} f_n} \qquad (7.14)$$

When a food source is not profitable after a *limit* number of cycles that food source is abandoned. Now, scout bees are sent out in search for new solutions that enable exploration. The best food source found by all the bees is output as the solution to the problem on completion of maximum iterations.

ABC-Based Task Scheduling Algorithm is given in Algorithm 2. Steps 2–4 of the algorithm are used to generate bee positions randomly, satisfying the constraints mentioned in the mathematical model, and their nectar amounts are also calculated. Steps 8–16 find the new position and the corresponding nectar amount for each employed bee. For example, if (5, 4, 1, 3, 2) is the current solution x_i and if x_k is (4,1,2,3,5) and $\phi_i = 0.5$, then $v_{ij} = (5.5, 5.5, 0.5, 3, 0.5)$, which is converted to discrete values (1, 1, 1, 3, 1) with $m = 5$ using the equations (7.12) and (7.13) from which duplicate entries are removed to get (1, 2, 4, 3, 5). The nectar amount Z (Equation (7.6)) of each of these bees is computed for the given θ value, and the position is updated if it finds to be better.

In steps 18–31, onlooker bees compute the probability of solution fitness considering its own fitness and fitness of all the bees in the swarm. If the computed value is higher than a predefined value, the position is updated else it will retain its previous position. In every iteration of the algorithm (limit = 1), a scout bee is the one that stores the solution with the maximum Z value. A new scout solution is then generated by randomly interchanging one or two (Task, VM) assignments from the current scout solution that helps

Algorithm 2 ABC-Based Task Scheduling Algorithm

1: Initialize E, U, pb, θ, max, ϕ, $n = 1$, $i = 1$
2: $EB(n) \leftarrow$ Generate E random schedules
3: $OB(n) \leftarrow$ Generate U random schedules
4: Calculate M, C and $Z(n)$ for each schedule /as in Equations (7.1), (7.2), (7.6)
5: $n = n + 1$
6: **while** $(n <= \text{max})$ **do**
7:   $i = 1$, $Tf = 0$
8:   **for** (each $EB(n)(i), i \leq E$) **do**
9:     $k = \text{random}(E)$
10:     $EB(n)(i) = EB(n-1)(i) + \phi * EB(n-1)(k)$
11:     $EB(n)(i) \leftarrow$ Apply Equations (7.12) and (7.13) on $(EB(n)(i))$
12:     Calculate $M(i)$, $C(i)$ and $ZEB(n)(i)$
13:     **if** $(ZEB(n)(i) > ZEB(n-1)(i))$ **then**
14:       $EB(n)(i) = EB(n-1)(i)$
15:     **end if**
16:   **end for**
17:   Sort $EB(n)$ based on $ZEB(n)$
18:   **for** $(i = 1$ to E$)$ **do**
19:     $Tf = Tf + ZEB(n)(i)$
20:   **end for**
21:   **for** $(i = 1$ to U$)$ **do**
22:     $Tf = Tf + ZOB(n-1)(i)$
23:   **end for**
24:   **for** (each $OB(n)(i), i \leq U$) **do**
25:     $p = ZOB(n-1)(i)/Tf$
26:     **if** $(p \geq pb)$ **then**
27:       $OB(n)(i) = EB(n)(i)$
28:     **else**
29:       $OB(n)(i) = OB(n-1)(i)$
30:     **end if**
31:   **end for**
32:   Scout$(n) \leftarrow EB(n)$ or $OB(n)$ with maximum Z
33:   Scout$(n) \leftarrow$ Interchange two positions of Scout(n) randomly
34:   Calculate M, C and Z of $Scout(n)$
35:   Replace corresponding $EB(n)$ or $OB(n)$ with $Scout(n)$, $n = n + 1$
36: **end while**
37: Output Best among $EB(n)$ and $OB(n)$

for exploration, and the fitness value is calculated (steps 32–35). For example, if the scout bee is identified to be (5,4,1,3,2), then exchanging 5 with 1 (position 1 with position 3) yields a new solution (1,4,5,3,2), which is added to the solution set. Steps 6–36 are repeated for *max* iterations, and on completion, the algorithm outputs the best bee position among all the bees as the solution to the optimization problem.

7.4.3   Genetic Algorithm

GA is based on genetic science and natural selection. It is also a population-based meta-heuristic search technique. It works under the principle of *'survival of the fittest'*. The chance of being selected as the parent of the next generation depends upon the fitness value of individuals. At each iteration, selected parent solutions are recombined (crossover operation) to generate new off-spring solutions. The mutation operator is then applied to a very small set of offspring solutions. The new solutions replace other solutions according to a replacement strategy.

Task scheduling using GA is given as Algorithm 3. A solution or a chromosome is composed of a sequence of n genes, where each gene represents a task to VM assignment, represented using permutation encoding. Initial solutions are randomly generated meeting constraints, and their fitness is evaluated (steps 2–3). Steps 6–19 of the algorithm select parent individuals for generating offsprings for the next generation probabilistically through Roulette wheel selection method. The scaling method applied here is exponential scaling. The selected chromosomes are recombined to produce offsprings. The recombination procedure adopted is one point crossover technique on 90% of these individuals, where the crossover point is determined randomly on each generation (steps 21–23). For example, if the point of crossover selected is 2 and if the parent chromosomes are (5,4,1,3,2) and (4,1,2,3,5), then new off-springs are (5,4,2,3,5) and (4,1,1,3,2). The offsprings after duplicate removal are (5,4,2,3,1) and (4,1,5,3,2). Mutation operation is performed on a small set of offspring chromosomes (5%) for exploitation. The mutation strategy adopted is an interchange of two randomly selected genes (step 24). For example, if (5,4,1,3,2) is the offspring chromosome that is selected for mutation, two positions are randomly selected, say 2 and 4, then the chromosome after mutation will be (5,3,1,4,2). These offspring chromosomes are evaluated to find the value of Z (*fitness*) for the given θ value. Steps 25–27 of the algorithm removes duplicates and find the fitness value of each offspring chromosome. The replacement strategy adopted is generational elitism, where the best N solutions (solutions with a minimal Z value) of parent and offspring chromosomes are selected for the next generation (step 28). Steps 4–29 are repeated for *max* generations, and on completion, the best chromosome is output as the solution to the problem. An initial proposal on the above algorithm is presented in our earlier work [3].

7.4.4   Ant Colony Optimization Algorithm

ACO algorithm is a construction-based search algorithm for an optimization problem. They use the foraging behavior of ant colonies. Ants deposit a special

---

Algorithm 3 GA-Based Task Scheduling

1: Initialize N, K, θ, max, G, $n = 1$, $i = 1$
2: $P(n) \leftarrow$ Generate N random schedule
3: Calculate M, C and $Z(n)$ for $P(n)$ /as in Equations (7.1), (7.2), (7.6)
4: **while** $(n <= \max)$ **do**
5:    sum= 0
6:    **for** (each $P(n)(i), i \leq N$) **do**
7:      $sc(i) = exp(Z(n)(i) * K)$, sum = sum $+ sc(i)$
8:    **end for**
9:    **for** $(i = 1$ to N) **do**
10:      $pb(i) = sc(i)/$sum
11:    **end for**
12:    **for** $(i = 2$ to N) **do**
13:      $cpb(i) = cpb(i - 1) + pb(i)$ with $cpb(1) = pb(1)$
14:    **end for**
15:    **for** $(k = 1$ to N) **do**
16:      **for** $(i = 1$ to N) **do**
17:        $Pt(n)(k) \leftarrow P(n)(i)$ depending on random(N) & $cpb(i)$
18:      **end for**
19:    **end for**
20:    **for** $(k = 1$ to N, k = k+2) **do**
21:      $i = 1$, $j = $ random(N)
22:      Do crossover at j on $Pt(n)(k)$ & $Pt(n)(k+1)$ to get $ch(k)$ & $ch(k+1)$
23:    **end for**
24:    $j = $ random(G), $k = $ random(G), Interchange $(ch(N : j), ch(N : k))$
25:    **for** (each $ch(i)$, $i \leq N$) **do**
26:      Remove duplicates from $ch(i)$, Calculate $M(i)$, $C(i)$ and $Z(i)$
27:    **end for**
28:    $P \leftarrow$ Best N among $P(n)$ and ch based on Z, $n = n + 1$
29: **end while**
30: Output P

---

kind of chemical substance called pheromones on the ground while moving towards a food source. Ants construct solutions, and these are used to perform a pheromone update to modify the pheromone trails. The pheromone trails affect the construction of solutions in the next iteration. Algorithm for task scheduling based on ACO technique is given as Algorithm 4. During a tour, the ants build the solution to the cloud task scheduling problem by adding one VM from a set of VMs for a particular task, and the process repeats until all tasks are assigned some VMs, thus completing a tour. Initially, one VM is assigned randomly for each ant (steps 3–5). Then, the selection of a particular VM for a given task is determined by a selection probability (steps 6–12 of the algorithm). Let τ_{ij} be the pheromone concentration on the path from $Task_i$ to VM_j and η_{ij} be the heuristic information associated on the same path, then

---

Algorithm 4 ACO-Based Task Scheduling Algorithm

1: Initialize N_a, α, β, θ, ρ, Q N, max, $k = 1$ $S = NULL$
2: **while** $(k < max)$ **do**
3:     **for** $(i = 1$ to $N_a)$ **do**
4:         $S(i) \leftarrow$ Assign 1 VM randomly
5:     **end for**
6:     $t = 2$
7:     **while** $(t \leq N)$ **do**
8:         **for** (each ant_i, $i = 1$ to N_a) **do**
9:             $S(i) \leftarrow$ Choose next VM using Equation (7.15)
10:        **end for**
11:        $t = t + 1$
12:    **end while**
13:    Update pheromone trails using Equation (7.16)
14:    Apply pheromone evaporation using Equation (7.17)
15:    **for** ($i = 1$ to N_a) **do**
16:        Calculate $M(i)$, $C(i)$ and $Z(i)$ of $S(i)$
17:    **end for**
18:    $S^* \leftarrow$ Best solution from S
19:    Update global pheromone evaporation using Equation (7.18)
20:    $k = k + 1$
21: **end while**
22: Return S^*

---

the probability of selecting j^{th} VM for $Task_i$ for ant k is determined using the following formula:

$$p_{ij}^k = \frac{[\tau_{ij}]^\alpha * [\eta_{ij}]^\beta}{\sum_{l \in N_i} [\tau_{il}]^\alpha * [\eta_{il}]^\beta} \tag{7.15}$$

where α and β are parameters that determine the impact of pheromones and heuristic information, respectively, and l is the set of allowed VMs for ant k in the current tour. The inverse of objective function value is taken as heuristic information (η). With the selection of a particular path, each ant lays a quantity of pheromone ($\Delta \tau_{ij}$) on that path, determined by Equation (7.16) where Q is an adaptive parameter and Z_{ij} is the objective function value for $Task_i$ on VM_j (step 13). In step 14, the pheromone update mechanism defined by Equation (7.17) is done where ρ is a parameter to control pheromone evaporation.

$$\Delta \tau_{ij} = \frac{Q}{Z_{ij}} \tag{7.16}$$

$$\tau_{ij} = (1 - \rho) * \tau_{ij} + \Delta \tau_{ij} \tag{7.17}$$

When all ants complete the solution construction process, the fitness of the solutions is evaluated and the best-found solution is updated from among all the ants (steps 15–18). The τ_{ij} for the best solution path is again updated with a global perspective, using Equation (7.18), which helps to bias or broaden the search in the next iteration (step 19). The algorithm iterates for *max* number of times (steps 7–21) and on completion returns the best solution.

$$\Delta \tau_{ij} = \frac{Q}{Z_{\text{best}}} \tag{7.18}$$

7.5   Experimental Settings

Three types of synthetic data sets, *Type 1*, *Type 2* and *Type 3*, are used to analyze the performance of the above four algorithms. In *Type 1* data set, a task set of 99 tasks is randomly generated with a task length in the range of 1–9, and 9 VMs of different capability levels are taken for scheduling. *Type 2* data sets and *Type 3* data sets are generated according to the procedure for Expected Time to Compute (ETC) matrix described by Braun et al. [4]. *Type 2* data sets have the property that the variability in task length exhibits random probability distribution and *Type 3* data sets have task length variability of normal probability distribution. Each ith entry represents the estimated execution time of $task_i$ on a 'small'-type VM. The characteristics of the ETC matrix is varied to represent task heterogeneity ($\phi_b = 100, 1,000, 3,000$) and machine heterogeneity ($\phi_r = 10, 100, 1,000$). Sample 4×4 excerpt from one of the 512 ETC matrices is shown in Table 7.2. Experiments are done with 9 ETC matrices of size 512, another 9 of size 1024 and another 9 having 2048 entries. For an ETC matrix of 512 entries, 16 VMs are considered, and for 1024 and 2048 tasks, 32 VMs of different capability levels are considered. For comparing the performance of the biologically inspired algorithms, another meta-heuristic algorithm, namely, Greedy Randomized Adaptive Search Procedure (GRASP) is also simulated. Each of these five algorithms is trying to minimize the value of the M function (Equation (7.1)) and C function (Equation (7.2)) simultaneously, so as to have a minimum value of Z (Equation (7.6)) satisfying constraint equations presented in Section 7.3.

Type 1 and *Type 2* data sets have sets of tasks showing random probability distribution. These data sets are used to characterize cloud computing systems

TABLE 7.2
Sample 4×4 *Type 2* Data Set

167177.182	19862.324	27226.802	116593.838
75809.844	80844.535	2957.051	69479.025
65871.960	970.218	78680.239	34823.485
103666.221	191988.413	51695.280	66.433

where the task arrival pattern is dynamic and unpredictable. For most of the days or hours, the behavior of the system is purely random. *Type 3* data set tasks are generated to characterize bursty traffic in the cloud computing systems, where there will be a huge demand for certain applications running in the cloud, giving spikes. For this period, the task arrival rate and task length show a bell-shaped curve. Also, it is reported in the literature that Map-Reduce jobs in the cloud can be better characterized through log-normal probability distribution. Hence, the data distribution in *Type 3* data set is chosen to be of a normal probability distribution. The probability distribution of the sample data sets for *Type 2* (random) and *Type 3* (normal) data sets is shown in Figures 7.1a and b, respectively.

Initial solutions are randomly generated for PSO, ABC, and GA techniques. In the solution, it is ensured that every VM is assigned only one task from the task row and more than one task is not assigned to the same VM in an iteration. Experiments are done by varying θ from 0 to 1 at an interval of 0.1, on *Type 1* data sets to study the behavior of each of the algorithms. When $\theta = 0$, the optimization algorithm finds running time optimal schedule, and when $\theta = 1$, the algorithm determines cost optimal schedule which is equivalent to finding the solution for the single-objective optimization problem. Each of the algorithms is tested on *Type 1* data sets for a different number of tasks, a different number of VMs and different types of VMs. For further analysis, *Type 2* data sets and *Type 3* data sets are used. The termination criteria for all algorithms are set as 500 generations, which is determined from the convergence study. Each of these algorithms is run ten times, and the average values of *cost* and *makespan* are presented. The parameters needed for each of the algorithms are also set as given below:

PSO Settings: The inertia weight $w = 0.6$, $c_1 = c_2 = 0.2$, $v_{min} = -4.0$, $v_{max} = 4.0$ and $N = 25$.

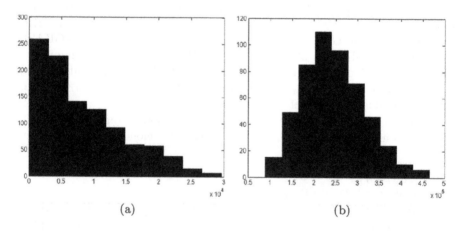

(a)   (b)

FIGURE 7.1
Probability distribution of data sets. (a) Type 1 & Type 2, (b) Type 3.

ABC Settings: The number of food sources $N = 25$, the number of employed bees $E = 13$, number of onlooker bees $U = 12$ and limit $= 1$.

GA Settings: The population size $N = 25$. The crossover probability is 90% and the mutation probability is 5%.

ACO Settings: Initial pheromone value $\tau = 0.5$, $N_a = 25$, $Q = 1/10$, $\alpha = 0.3$, $\beta = 1$ and $\rho = 0.4$.

7.6 Results and Discussion

Figure 7.2a represents the average convergence values on cost for different values of θ, and Figure 7.2b is that of *makespan* on the *Type 1* data set. In both these figures, it is seen that the values obtained by the GRASP and ACO algorithms are much higher when compared to GA, PSO, and ABC algorithms for different values of θ. Hence, for further analysis on the *Type 1* data set, only the performances of GA, PSO, and ABC algorithms are compared. The convergence characteristics of these algorithms for $\theta = 0.9$ are further studied. For this θ value, the PSO algorithm finds an optimal cost value in 90% of the time, but both GA and ABC algorithms have not converged to the optimal cost value on multiple trials.

A detailed performance analysis of these algorithms is done by varying the number of tasks from 18 to 99 (refer Figures 7.3a and b), with a different number of VMs from 6 to 9 varying the number of task rows from 17 to 11 with appropriate padding (refer Figures 7.4a and b) and various types of VMs (refer Figures 7.5a and b) on the *Type 1* data set. The average results of 10 independent trials are presented in these figures. It can be seen that the

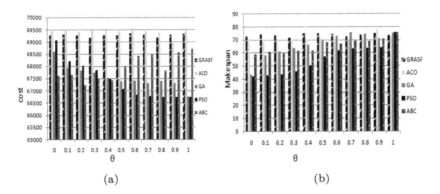

(a)　　　　　　　　　　　　(b)

FIGURE 7.2
Pareto-optimal solutions - *Type 1* data sets. (a) Cost, (b) Makespan.

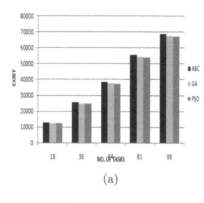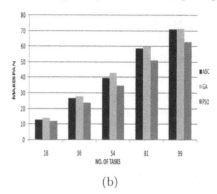

FIGURE 7.3
Type 1 data set - Different number of Tasks (a) Cost Vs Tasks, (b) Makespan Vs Tasks.

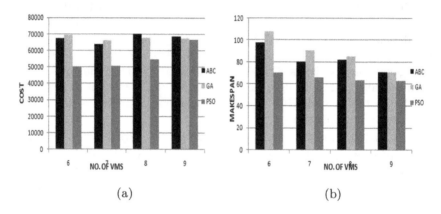

FIGURE 7.4
Type 1 data set - different number of VMs. (a) Cost Vs VMs, (b) Makespan Vs VMs.

performance of PSO algorithm is better than the GA, and both are better than the ABC algorithm for most of the cases.

To study the performance of these algorithms on a more number of tasks and VM pairs, all these algorithms are executed on nine, *Type 2* data sets having a random probability distribution and on nine, *Type 3* data sets having normal probability distribution on task length, characterizing regular traffic and bursty traffic, respectively. The average *cost* and *makespan* of each of these algorithms are compared against GRASP algorithm, and the average performance improvement in % of 10 independent trials is presented in Table 7.3. The % performance benefit of *Algorithm X* on factor a (denoted as X_a) with respect to *Algorithm Y* on the same factor (denoted as Y_a) is calculated using the following equation:

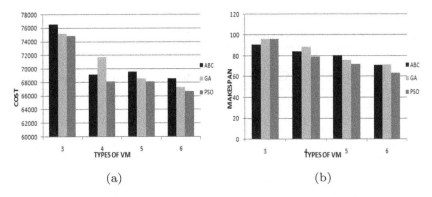

(a)   (b)

FIGURE 7.5
Type 1 data set - different types of VMs. (a) Cost Vs VM Types, (b) Makespan
Vs VM Types.

TABLE 7.3
Result Comparison with GRASP

Algorithm	Type 2		Type 3	
	Cost %	Makespan %	Cost %	Makespan %
ACO	0.102	−4.212	−0.019	−1.812
ABC	1.491	16.687	0.485	7.827
GA	5.751	31.281	1.657	13.923
PSO	9.779	57.356	3.272	23.732

$$a\% = \frac{(Y_a - X_a)}{Y_a} * 100 \qquad (7.19)$$

In Table 7.3, it is seen that the performance of ACO algorithm is
poor when compared to GRASP algorithm. All the other algorithms are
better than GRASP in both the objective function values. Also, both
GA and PSO algorithm produce better results than the other three
algorithms.

The three better-performing algorithms (GA, ABC, and PSO) are again
analyzed for *Type 2* and *Type 3* data sets having a more number of tasks with
double the number of VMs. The performance of PSO algorithm and GA is
compared against the ABC algorithm, and the average % improvement in *cost*
and *makespan* is shown in Table 7.4 for $\theta = 0.9$. We have considered nine,
1024 task set and nine, 2048 task set with 32 VMs each, and their average
results are presented. As it is evident from the table, the % benefit of the
PSO algorithm is far better than GA for any data set types. Also, there are
no significant performance benefits among Type 2 or Type 3 data sets for
PSO algorithm or GA implying that these algorithms perform similarly on
any type of traffic in the cloud systems.

TABLE 7.4

Result Comparison with ABC Algorithm

		Type 2		Type 3	
Algorithm	Data Set	Cost %	Makespan %	Cost %	Makespan %
GA	1024	2.145	9.055	0.661	3.478
	2048	2.129	8.623	0.708	3.295
PSO	1024	6.68	36.620	2.130	13.853
	2048	6.718	35.963	2.163	14.004

The performance of each of these algorithms is analyzed using two quality parameter measures, namely the *Degree of Imbalance* (DOI) [6,12] and *Standard Deviation* (σ) [6,19] which are defined in Equations (7.20) and (7.21), respectively. DOI is a measure to know the load balancing ability among heterogeneous VMs. A small value of DOI is preferred. *Standard Deviation* on task completion time is an indication of the search efficiency of the algorithm in finding optimal *makespan*. A small value of σ is an indication of the algorithm's effectiveness in finding stable solutions.

$$\text{DOI} = \frac{(T_{\max} - T_{\min})}{T_{\text{avg}}} \tag{7.20}$$

$$\sigma = \sqrt{\frac{1}{N} \sum_{i=1}^{N} (x_i - \bar{x})^2} \tag{7.21}$$

where $T_{\max}$, $T_{\min}$ and T_{avg} are the maximum, minimum, and average task lengths T_i, respectively, submitted among all VMs in an epoch; x_i is the task completion time of VM_i; and $\bar{x}$ is the mean task completion time of all VMs.

Tables 7.5 and 7.6 show a comparison of *DOI* and σ for different algorithms. It can be seen that the load of VMs is more balanced in PSO technique as they have the small DOI value and comparatively better in GA than in other algorithms. Similarly, the value of σ is comparatively small for PSO algorithm, showing its capacity to search better in the objective space.

From the analysis, it can be concluded that the GA and PSO algorithm are comparatively better for task scheduling in the cloud computing environment, irrespective of the data characteristics. ACO and GRASP algorithms are unable to achieve an optimal value in any of the cases studied. The ABC algorithm performs better than both GRASP and ACO algorithms, but its

TABLE 7.5

DOI & σ for Type 1 Data Sets

	GRASP	ACO	ABC	GA	PSO
DOI	2.054	2.181	2.241	2.06	1.809
σ	1.792	2.082	1.912	1.722	1.32

TABLE 7.6
DOI & σ for Type 2 Data Sets

Data Set		ABC	GA	PSO
512	DOI	3.543	3.469	2.439
	σ	321.2	297.2	159.9
1024	DOI	4.123	4.236	3.2
	σ	115.6	109.36	71.688
2048	DOI	4.174	4.24	3.31
	σ	119.9	107.52	72.668

performance is poor when compared to GA and PSO techniques. The PSO technique outperforms all the other algorithms in any of the aspects considered. The running time needed for the convergence of PSO techniques is also less when compared to other bio-inspired algorithms.

7.7 Conclusion

Scheduling tasks in the cloud is a challenging process as it involves many factors such as cost, energy-efficient management and profit considerations at CSP's end, and execution time and other QoS parameters at the user's end. In the cloud environment, the task arrival pattern is also highly unpredictable and dynamic in nature. For scheduling tasks in this complex environment, the problem is modeled as a constraint bi-objective optimization problem, where the objectives considered are *makespan* and *cost* and have used five metaheuristic search algorithms. A weighted sum–based multi-objective optimization approach is employed to find the optimal solutions. The experimental result shows that the performance of PSO technique is far better than other bio-inspired optimization algorithms irrespective of the task arrival patterns and length of tasks. This study provides a comparison and gives an insight into the suitability of each of these techniques in the cloud domain. As a future work, the performance of these algorithms in the scheduling of workflow applications needs to be explored. Similarly other methods to find pareto-optimal solutions are to be experimented.

Bibliography

[1] E.S. Alkayal, N.R. Jennings, and M.F. Abulkhair. Efficient task scheduling multi-objective particle swarm optimization in cloud computing. In *Proceedings of 41st Conference on Local Computer Networks Workshops*, Dubai, United Arab Emirates, pages 17–24. IEEE, 2016.

[2] A.S.A. Beegom and M.S. Rajasree. A particle swarm optimization based pareto-optimal task scheduling in cloud computing. *Lecture Notes in Computer Science*, 8795:79–86, 2014.

[3] A.S.A. Beegom and M.S. Rajasree. Genetic Algorithm framework for bi-objective task scheduling in cloud computing systems. *Lecture Notes in Computer Science*, 8956:356–359, 2015.

[4] T.D. Braun, H.J. Seigel, N. Beck, L.L. Boloni, M. Maheswaran, A.I. Reuther, J.P. Robertson, M.D. Theys, B. Yao, D. Hensegen, and R.F. Freund. A comparison of eleven static heuristics for mapping a class of independent tasks onto heterogeneous distributed computing systems. *Journal of Parallel and Distributed Computing*, 61:810–837, 2001.

[5] Q. Cao, Z. Wei, and W. Gong. An optimized algorithm for task scheduling based on activity based costing in cloud computing. In *Proceedings of 3rd International Conference on Bioinformatics and Biomedical Engineering (ICBBE 2009)*, Beijing, China, pages 1–3. IEEE, 2009.

[6] G.F. Elhady and M.A. Tawfeek. A comparative study into swarm intelligence algorithms for dynamic task scheduling in cloud computing. In *Proceedings of 7th International Conference on Intelligent Computing and Information Systems*, FeCairo, Egypt, pages 362–369. IEEE, 2015.

[7] M. Feng, X. Wang, Y. Zhang, and J. Li. Multi-objective particle swarm optimization for resource allocation in cloud computing. In *Proceedings of 2nd International Conference on Cloud Computing and Intelligent Systems (CCIS)*, Hangzhou, China, pages 1161–1165. Springer, 2012.

[8] L. Guo, G. Shao, and S. Zhao. Multi-objective task assignment in cloud computing by particle swarm optimization. In *Proceedings of 8th International Conference on Wireless Communications, Networking and Mobile Computing (WiCOM)*, Shanghai, China, pages 1–4. IEEE, 2012.

[9] J. Jin, J. Luo, A. Song, F. Dong, and R. Xiong. BAR : An efficient data locality driven task scheduling algorithm for cloud computing. In *Proceedings of 11th IEEE/ACM International Symposium on Cluster, Cloud and Grid Computing*, Newport Beach, CA, USA, pages 295–304. IEEE/ACM, 2011.

[10] G. Lee. Resource allocation and scheduling in heterogeneous cloud environments. PhD Thesis report of Department of Electrical Engineering and Computer Science, University of California, Berkeley, 2012.

[11] V.A. Leena, A.S.A. Beegom, and M.S. Rajasree. Genetic algorithm based bi-objective task scheduling in hybrid cloud platform. *International Journal of Computer Theory and Engineering*, 8:7–13, 2016.

[12] K. Li, G. Xu, G. Zhao, Y. Dong, and D. Wang. Cloud task scheduling based on load balancing ant colony optimization. In *Proceedings of Sixth IEEE Annual ChinaGrid Conference*, Liaoning, China, pages 3–9. IEEE, 2011.

[13] R.T. Marler and J.S. Arora. The weighted sum method for multi-objective optimization : new insights. *Structural and Multidisciplinary Optimization*, 41:853–862, 2010.

[14] S. Selvarani and G.S. Sadhasivam. Improved cost-based algorithm for task scheduling in cloud computing. In *Proceedings of IEEE International Conference on Computational Intelligence and Computing Research (ICCIC)*, Coimbatore, India, pages 1–5. IEEE, 2010.

[15] I.P. Stanimirovic, M.L. Zlatanovic, and M.D. Petkovic. On the linear weighted sum method for multi-objective optimization. *Facta Universitatis, Series: Mathematics and Informatics*, 26:49–63, 2011.

[16] C. Tsai, W. Huang, M.H. Chiang, M.C. Chiang, and C. Yang. A hyper-heuristic scheduling algorithm for cloud. *IEEE Transactions on Cloud Computing*, 2:236–250, 2014.

[17] X. Wang and Y. Wang. An energy and data locality aware bi-level multiobjective task scheduling model based on MapReduce for cloud computing. In *Proceedings of IEEE/WIC/ACM International Conference on Web Intelligence and Intelligent Agent Technology*, Macau, China, pages 648–655. IEEE/ACM, 2012.

[18] Z. Zhan, X. Liu, Y. Gong, J. Zhang, H.S. Chung, and Y. Li. Cloud computing resource scheduling and a survey of its evolutionary approaches. *ACM Computing Surveys*, 47:1–33, 2015.

[19] X. Zuo, G. Zhang, and W. Tan. Self-adaptive learning PSO based deadline constrained task scheduling for hybrid IaaS cloud. *IEEE Transactions on Automation Science and Engineering*, 11(12):564–573, 2014.

Index

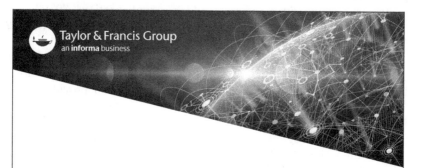

Taylor & Francis eBooks

www.taylorfrancis.com

A single destination for eBooks from Taylor & Francis
with increased functionality and an improved user
experience to meet the needs of our customers.

90,000+ eBooks of award-winning academic content in
Humanities, Social Science, Science, Technology, Engineering,
and Medical written by a global network of editors and authors.

TAYLOR & FRANCIS EBOOKS OFFERS:

A streamlined
experience for
our library
customers

A single point
of discovery
for all of our
eBook content

Improved
search and
discovery of
content at both
book and
chapter level

REQUEST A FREE TRIAL
support@taylorfrancis.com